Fiber Optic Communications

Fiber Optic Communications

HAROLD B. KILLEN
College of Technology
University of Houston

PRENTICE HALL, Englewood Cliffs, New Jersey 07632

ST. PHILIP'S COLLEGE LIBRARY

Library of Congress Cataloging-in-Publication Data

Killen, Harold B.
   Fiber optic communications / Harold B. Killen.
      p. cm.
   Includes index.
   ISBN 0-13-313578-0
   1. Optical communications.   2. Fiber optics.   I. Title.
TK5103.59.K54 1991
621.382′75—dc20
                                       90-25700
                                          CIP

Acquisitions editor: *Sharon Jacobus*
Editorial/production supervision: *Raeia Maes*
Manufacturing buyers: *Mary McCartney/Ed O'Dougherty*
Cover design: *Ben Santora*

 © 1991 by Prentice-Hall, Inc.
A division of Simon & Schuster
Englewood Cliffs, New Jersey 07632

All rights reserved. No part of this book may be
reproduced, in any form or by any means,
without permission in writing from the publisher.

Printed in the United States of America

10 9 8 7 6 5 4 3 2 1

ISBN 0-13-313578-0

Prentice-Hall International (UK) Limited, *London*
Prentice-Hall of Australia Pty. Limited, *Sydney*
Prentice-Hall Canada Inc., *Toronto*
Prentice-Hall Hispanoamericana, S.A., *Mexico*
Prentice-Hall of India Private Limited, *New Delhi*
Prentice-Hall of Japan, Inc., *Tokyo*
Simon & Schuster Asia Pte. Ltd., *Singapore*
Editora Prentice-Hall do Brasil, Ltda., *Rio de Janeiro*

To my wife
PHYLLIS JEANNE KILLEN

Contents

PREFACE xi

1 INTRODUCTION TO FIBER OPTICS 1

   1.0 Introduction, 1
   1.1 Optical Fiber Transmission Parameter, 2
       1.1.1 Dispersion, 2
       1.1.2 Single-Mode Fiber, 6
       1.1.3 Graded-Index Fiber, 9
       1.1.4 Impulse Response, 10
       1.1.5 System Losses, 12
   1.2 Detectors and Sources, 15
       1.2.1 Source Launch Profile, 16
       1.2.2 Reflection Loss, 18
       1.2.3 Source Requirements, 19
       1.2.4 Detector Requirements, 25
   1.3 System Design Parameters, 29
       1.3.1 Component Selection, 29
       1.3.2 System Architecture, 30
       1.3.3 Signal Multiplexing, 32
       1.3.4 Modulation, 33

   1.3.5 Signal-to-Noise Ratio, 36
   1.3.6 Digital Modulation, 36
   1.3.7 Loss Budgeting in System Design, 37
   1.3.8 Signal Degradation, 39
 1.4 Drive and Receive Circuits, 40
   1.4.1 Photodiode Receiver Circuits, 40

2 ANALOG MODULATION   43

 2.0 Introduction, 43
 2.1 Optical Channel, 43
   2.1.1 Detection of Optical Radiation, 44
   2.1.2 Direction Detection Optical Communication Receiver, 46
 2.2 Baseband Analog Modulation and Detection, 51
   2.2.1 ADP Detector, 54
 2.3 Multiple-Channel Modulation, 57
   2.3.1 Analog Modulation Methods, 59
 2.4 Design of Frequency Division Multiplexed Fiber Optic Links, 61
   2.4.1 Key System Performance Parameters, 63
   2.4.2 Fiber/Receiver Carrier-to-Noise Contribution, 65
   2.4.3 Bandwidth Limitations, 67
   2.4.4 Distortion, 68
   2.4.5 Alternative Design Approaches, 70
 2.5 Pulse Modulation, 72
   2.5.1 System Rise Time, 72
   2.5.2 Pulse Frequency Modulation, 73
   2.5.3 Multichannel PFM Video (FDM), 75

3 DIGITAL FIBER OPTIC SYSTEM DESIGN   79

 3.0 Introduction, 79
 3.1 Digital System Peformance Parameters, 80
   3.1.1 Dispersion Constraints on Bit Rate, 84
 3.2 Designing Digital Fiber Optic Links, 86
   3.2.1 Distance and Data Rate, 89
   3.2.2 Selecting Components, 91
   3.2.3 Power Budget, 91
   3.2.4 Bandwidth Budget, 96
   3.2.5 Optical Transmitter and Receiver Considerations, 100
 3.3 Testing the Design, 103
 3.4 Data Bus Topology, 104

4 BASEBAND CODING FOR FIBER OPTICS   114

 4.0 Introduction, 114
 4.1 Source and Channel Coding Theorem, 115

4.2 Channel Encoding, 119
   4.2.1 Coding Principles, 120
4.3 Bandwidth of Digital Data, 121
4.4 Digital Signaling Techniques, 125
   4.4.1 Nonreturn to Zero (NRZ), 128
   4.4.2 Return to Zero (RZ), 128
   4.4.3 Biphase, 129
   4.4.4 Delay Modulation, 130
   4.4.5 Multilevel Binary, 131
4.5 Data Encoding for Fiber Optics, 131

5 DIGITAL VIDEO TRANSMISSION IN OPTICAL FIBER NETWORKS 143

5.0 Introduction, 143
5.1 Factors Affecting Digital Video, 144
   5.1.1 Digital Transmission System Hierarchy, 146
   5.1.2 NTSC Video, 146
5.2 Analog-to-Digital Conversion, 151
   5.2.1 Video Compression, 154
   5.2.2 Performance Requirements, 156
5.3 Designing Fiber Optic Networks for High Speed, 156

6 OPTICAL RECEIVERS 161

6.0 Introduction, 161
6.1 Optimization of Parameters in High-Speed Silicon Photodiodes, 164
   6.1.1 Charge-Collection Time, 164
   6.1.2 RC Rise Time Component, 165
   6.1.3 Diffusion Time, 167
   6.1.4 Total Photodiode Rise Time, 167
   6.1.5 Amplifier Design Considerations, 168
6.2 Optical Receivers, 171
   6.2.1 Noise Considerations, 173
   6.2.2 Critical Areas of Receiver Design, 176
   6.2.3 Optical Receiver Design Based on Nyquist's First and Second Criteria, 177

7 COHERENT OPTICAL COMMUNICATIONS 187

7.0 Introduction, 187
7.1 Coherent Transmission Techniques, 188
7.2 Modulation Techniques, 193
   7.2.1 Fiber Requirements, 194

7.3 Receiver Sensitivity, 196
    7.3.1 Fundamental Limits of Direct Detection, 196
    7.3.2 Homodyne Direct Detection and the Super Quantum Limit, 198
    7.3.3 Ideal Heterodyne Detection, 200
    7.3.4 Phase Noise in Lasers, 201
    7.3.5 Phase-Lock Techniques, 204

8 MEASUREMENTS IN FIBER TELECOMMUNICATIONS 210

8.0 Introduction, 210
8.1 Fiber Cable Measurements, 211
    8.1.1 Spectral Attenuation of Fibers, 211
    8.1.2 Optical Time-Domain Reflectometer, 213
    8.1.3 Dispersion Measurements, 214
8.2 Fiber Bandwidth Measurements, 217
    8.2.1 Comparison of Time- and Frequency-Domain Measurements, 218
    8.2.2 Bandwidth Specification by Fiber Manufacturers, 219
8.3 Optical Margin Measurements, 222
    8.3.1 Bit-Error-Rate Measurement, 223
8.4 Failure Types, 224

APPENDIX: RAY THEORY 227

INDEX 229

Preface

This is an introductory text intended to provide the practical knowledge needed to both understand and design today's rapidly evolving fiber optic systems. In general, the text is directed toward undergraduates, requiring for the most part a knowledge of algebra. It is designed to also appeal to practicing engineers and managers needing an introduction to the subject.

Chapter 1 presents an overview of fiber optic systems. It covers such topics as optical fiber transmission properties, optical sources, transmitters and receivers, analog and digital intensity modulation, and a preliminary introduction to link design (flux budgeting).

Chapter 2 presents the design and analysis of analog intensity modulation systems. This chapter discusses the optical channel and introduces the detection of optical radiation. Direct detection optical communication receivers are discussed, along with baseband analog modulation and detection techniques. Analysis of systems using both PIN and APD receivers is presented.

Chapter 3 covers digital fiber optic system design. The effects of dispersion on bit rate are introduced, and link design, including determination of the required optical power, is covered. Star and Tee couplers are introduced as part of the discussion of multiterminal fiber optic systems. Optical transmitters and receivers are discussed.

A discussion of baseband coding for fiber optics is presented in Chapter 4.

The classical digital system model is discussed, and the source and coding theorem is introduced. The bandwidth efficiency plane is introduced, along with a brief discussion of channel coding. This, in turn, leads to a discussion of bandwidth requirements for digital data. Digital signaling techniques and data encoding for fiber optics are discussed.

Chapter 5 presents a discussion of digital video transmission over optical fiber networks. NTSC video is discussed, along with the requirements for determining the sampling frequency of video. Bandwidth compression of video is introduced.

Optical receivers are introduced in several of the earlier chapters. Chapter 6 is devoted entirely to this subject. Optimization of the parameters in high-speed silicon photodiodes is discussed. R-C rise time is examined in conjunction with the design of the photodiode receiver. Critical areas of receiver design are discussed, as well as Nyquist's first and second criteria.

Coherent optical communications is introduced briefly in Chapter 2. Chapter 7 presents a discussion of coherent optical systems. Both homodyne and heterodyne systems are discussed. Laser phase noise is discussed along with phase locked techniques, and performance comparison is made with direct detection systems.

Chapter 8 discusses measurements in fiber telecommunication systems. Typical measurements are identified and measurement techniques are examined. Test equipment for performing system measurements are discussed, and applications such as dispersion and fiber bandwidth measurements are presented. A discussion of failure types with typical troubleshooting procedure is included. A discussion of the meaning of manufacturer's bandwidth specifications (as related to installed systems) is presented.

Fiber optic systems are constantly evolving, and a textbook can present only the basics of current material. To this end it is hoped that mastery of this material will prepare the student for further reading in the literature. End-of-chapter problems are included to test and extend the student's subject knowledge.

The author wishes to express his appreciation to the associates and reviewers who contributed to the development of this text. This also applies to organizations that gave permission to use technical material. Special thanks is due the Prentice Hall staff who produced the book.

Harold B. Killen

Fiber Optic Communications

1

Introduction to Fiber Optics

1.0 INTRODUCTION

The use of optical fiber as a replacement for copper media is proving to be an irresistible force for handling the enormous amounts of information that must be transmitted across the country or even around the world. Today it is also becoming the medium of choice for transmitting between and within buildings. The principles of guided light transmission were known in the nineteenth century. In 1870, John Tyndall demonstrated this principle before the British Royal Society. He showed that light was conducted in a curved path along an illuminated stream of water flowing from a hole in a tank. This experiment illustrated the concept of total internal reflection, wherein light rays propagate by reflection off the boundaries of a medium and escape primarily at the opposite end of the "conductor." A later researcher, Alexander Graham Bell, studied the possibility of transmitting speech on a beam of light to a device called a photophone. As early as 1910, theoretical studies had been completed by Hondros and Debye on dielectric waveguides.

The study of transmission by optical waves was continued in the 1920s and 1930s, but implementation of devices did not occur until the 1950s. During this period the flexible fiberscope, now widely used in medicine, was developed. In the 1960s an extensive investigation of cladded fiber waveguides was accomplished. During this period, the attenuation of the early fibers was on the order of 1000 dB/

km. Attention was focused on reducing these losses. By 1970, the attenuation was reduced to as low as 3 dB/km.

The next area of research by fiber manufacturers centered on reducing the dispersion (pulse-spreading) characteristics of the fiber. This feature limits the bandwidth and, therefore, the information capacity of the medium. These two factors—loss and dispersion—establish the distance between repeaters. The most powerful aspect of optical communication is the tremendous bandwidth available at optical frequencies. So far, this feature is the one least exploited. A wavelength of 1 μm corresponds to 300,000 GHz. Thus, a single 1-GHz channel corresponds to only 3.3 $\times$ 10^{-6} μm of wavelength spread. The future of lightwave communications is indeed bright.

In this chapter, we wish to introduce basic fiber optic concepts. Succeeding chapters will center around fiber optic technology, designing fiber optic links, and development and test of fiber optic systems. The concepts introduced will be reinforced with numerous solved examples.

1.1 OPTICAL FIBER TRANSMISSION PARAMETERS

When fiber optic communication systems are designed, several parameters enter into the cost-versus-performance calculations. Basically, these can be grouped into three categories: the light source and associated drive circuitry, the optical fiber and mechanical cable construction, and the photodetector and receiver circuitry. A firm understanding of these parameters is necessary in order to select the components making up a system. The performance limitations of an optical fiber system particularly impact total component cost. Bandwidth (dispersion) and optical power define the system's repeater spacing. Because each repeater requires a back-to-back photodetector/transmitter arrangement, it is desirable to either maximize repeater spacing or, better yet, eliminate the requirement completely. As noted in the introduction, dispersion and attenuation are the two parameters that govern the bandwidth and power limitations of an optical fiber system. These factors may be reduced by buying higher-priced fibers. From a systems viewpoint, optical power and bandwidth limits also depend on the light source (modulation rates, radiance, optical-coupling efficiency) and detector (responsivity, rise time, noise factors).

1.1.1 Dispersion

There are three basic fiber types: multimode step index, graded index, and single mode. Dispersion in a fiber varies with the fiber type. In discussing dispersion, the best optical parameter to start with is the index of refraction, n (App. A-1). We define the index of refraction for both glass and plastic as

$$n = \frac{c}{v} \qquad (1.1)$$

where

c = speed of light in a vacuum (3×10^8 m/sec)

v = speed of light in the fiber

The *index profile* for an optical fiber refers to how its refractive index varies as a function of radial distance from the center of the fiber. For example, in a fiber with a "step-index" profile, the refractive index exhibits an abrupt change (step) in value at radius, r_c; see Figure 1.1. From this figure, we see that the fiber's cross section divides into two regions: the circular central "core" and a surrounding annular "cladding."

In a step-index fiber, propagation of the optical energy occurs through total internal reflection at the core-cladding interface. Thus, the core index must be greater than the cladding index. Actually air, with an index of 1.0, could serve as the cladding. However, the cladding also serves as core support and cladding with an index slightly less than the core is chosen. We also see that the cladding-to-core index ratio is a prime factor in determining the dispersion of an optical pulse. Refer to Figure 1.1. Typical core and cladding indices are $n_1 = 1.48$ and $n_2 = 1.46$.

A rigorous analysis of energy propagation within an optical fiber involves the solution of Maxwell's (electromagnetic) equations. In common with metallic waveguides, this analysis shows that the propagating energy (light) is distributed among a discrete set of superimposed fields called *modes*. Differences in the propagation

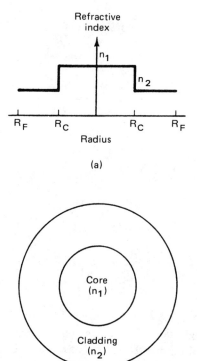

Figure 1.1 (a) The index profile takes a downward step at the core radius in a step-index fiber. (b) This step divides the fiber's cross section into a central core with index n_1 and a surrounding cladding with index n_2.

characteristics of these individual modes give rise to *modal dispersion*. This feature is one factor that limits the bandwidth of a fiber.

While the propagation characteristics of individual modes result in modal dispersion, it is common practice to use a geometrical ray approach to illustrate the approach. This technique is illustrated in Example 1.1.

Example 1.1

A 1-km fiber has core and cladding indices of $n_1 = 1.48$ and $n_2 = 1.46$. The core diameter is 50 µm. Find: (a) the minimum angle that supports total internal reflections, (b) the time delay of an off-axis ray compared to a ray that propagates directly down the fiber, and (c) an estimate of the modal-dispersion-limited bandwidth.

Solution

(a) Applying Snell's law (see Appendix), we have

$$\sin \gamma_{min} = \frac{n_2}{n_1} = \frac{1.46}{1.48}$$

which results in $\theta_{max} = (90° - \gamma_{min}) = 90° - 80.6° = 9.4°$.

(b)
$$\Delta t_{diff} = \frac{\text{distance}}{v}$$

From Figure 1.2,

$$h = D \tan \gamma_{min} = (50 \text{ µm}) \tan 80.6° = 306.13$$

$$\text{Number of } h\text{'s} = \frac{h}{\sin \gamma_{min}} = \frac{h}{\sin 80.6°} = \frac{1000 \text{ m}}{h}$$

$$= 3.311 \times 10^6 \text{ m}$$

Number of L's = number of h's

Zigzag path distance = (number of L's)(L)

$$= (3.311 \times 10^6)(306.13 = 10^{-6})$$

$$= 1013.6 \text{ m}$$

$$\Delta_{distance} = 1014 - 1000 = 14 \text{ m}$$

$$\Delta t_{diff} = \frac{14}{(3 \times 10^8)/1.48} = 69 \text{ ns}$$

(c)
$$BW \approx \frac{1}{t_{diff}} = \frac{1}{69 \times 10^{-19}} = 14.5 \text{ MHz}$$

Observe from Example 1.1 that the propagation times of a bouncing ray and a ray that propagates straight down the central axis of the fiber differ by 69 ns. The temporal delay (dispersion) in the arrival time of these two rays along with those of rays traveling intermediate paths produces bit smearing or intersymbol interference

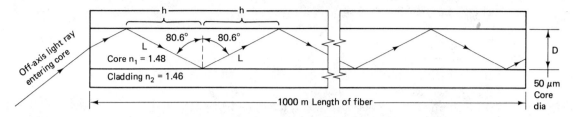

Figure 1.2 [From D.B. Keck, "Fundamentals of Optical Waveguide Fibers," *IEEE Communications Magazine*, Vol. 23, No. 5 (1985)].

in pulsed data systems and delay distortion in an analog modulated system. The dispersion limited bandwidth is approximately 15 MHz. In considering launch conditions of the light, it is obvious that it can enter the core-cladding interface at angles less than 80.6 deg. This energy propagates into the cladding and is lost.

We noted earlier that the energy actually propagates as superimposed fields called *modes*. Because of the numerous modes this fiber supports, it is called a *multimode step-index* fiber. From electromagnetic field theory, a mode volume parameter, *V*, may be expressed as

$$V = \frac{\pi d}{\lambda} \sqrt{n_1^2 - n_2^2} \qquad (1.2)$$

where

$$d = \text{fiber core diameter}$$
$$\lambda = \text{wavelength of the optical source}$$

Using Equation (1.1), we can estimate the number of propagating modes, *N*, in a step-index fiber as

$$N = \frac{V^2}{2} \qquad (1.3)$$

Example 1.2

Estimate the number of propagating modes for the fiber of Example 1.1. Assume $\lambda = 6 \times 10^{-2}$ m (red light).

Solution

$$V = \frac{\pi(50 \times 10^{-6})}{6 \times 10^{-7}} \sqrt{1.48^2 - 1.46^2} = 63.45$$

$$N = \frac{V^2}{2} = \frac{63.45^2}{2} = 2{,}012.95 \text{ modes}$$

1.1.2 Single-Mode Fiber

From Example 1.2, we see that the number of propagating modes is over 2000. Mode dispersion occurs in multimode fibers because different modes travel different effective distances through the fiber. The information-carrying capacity of a fiber is inversely related to the total dispersion. The total dispersion is a combination of three components: mode dispersion, material dispersion, and waveguide dispersion (see Section 1.1.2.2). It is clear from Equations (1.2) and (1.3) that the mode volume must be reduced to limit modal dispersion. Equation (1.2) suggests three ways for accomplishing this:

- Reduce the core diameter.
- Increase the wavelength.
- Decrease the difference between the core and cladding indices.

Now, absorption in the glass increases with an increase in wavelength. This effectively eliminates this option. We can make the ratio, n_1/n_2, as small as practicable. This ratio must, however, remain greater than one to maintain total internal reflection. The remaining approach is to reduce the core diameter. This makes coupling light into the small-diameter core more difficult. Advances in technology are in the process of solving this problem. For a single mode, $V = 1$, using Equation (1.2), we see that the diameter d of the fiber for a single mode is

$$d = \frac{V\lambda}{\pi\sqrt{n_1^2 - n_2^2}} = \frac{(1)(6 \times 10^{-7})}{\pi\sqrt{1.48^2 - 1.46^2}} = 7.88 \; \mu m$$

Studies indicate that if $V < 2.405$, only a single mode (an axial ray) can propagate.

We have seen that the size of the core is related to the ease of coupling light from the source to the fiber. For single-mode fibers, the light source should be a laser emitting in a single transverse mode. For a fiber with core diameter $2a$, a fractional index difference between core and cladding $\Delta = \Delta n/n$, and a homogenous core refractive index, a single mode will propagate for all wavelengths greater than the cutoff wavelength, λ_c, where

$$\lambda_c = \frac{2\pi an\sqrt{2\Delta}}{2.408} \tag{1.4}$$

and $\Delta = (n_1 - n_2)/n_1$. Additional modes will propagate for shorter wavelengths. Several single-mode designs that have recently been investigated are illustrated in Figure 1.3. In a single-mode fiber, no dispersion between the modes can exist and very high bandwidths are possible. The bandwidth of the single-mode fiber, however, is not infinite because of the other two sources of dispersion mentioned earlier—material and waveguide dispersion.

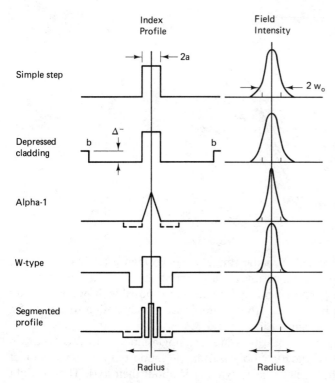

Figure 1.3 Radial refractive index profiles and light field intensities for several special single-mode fibers under investigation. These designs alter the dispersion characteristics of the single-mode fiber. [From D.B. Keck, "Fundamentals of Optical Waveguide Fibers," *IEEE Communications Magazine,* Vol. 23, No. 5 (1985)].

1.1.2.1 Emerging types of single-mode fibers. Present single-mode fibers permit transmission of data rates greater than 400 Mbits/s with repeater spacing in the tens of kilometers. There is widespread interest, however, for transmitting at even higher data rates over long distances. This is necessary to accommodate the growing volume of telecommunications both on land and between continents. This interest, along with special-purpose applications such as sensing, has led to new types of single-mode fibers.

The zero dispersion wavelength is 1.3 μm and the minimum loss wavelength is 1.5 μm (see Section 1.1.2.2). High data rates are limited by dispersion, while low data rates are limited by loss. Since dispersion is proportional to source spectral width, the optimization process consists of both trying to lower the attenuation at 1.55 μm or trying to narrow the laser spectral width. Attempts to narrow the spectral width have led to development of the distribution feedback laser. A grating in the semiconductor material reflects light of a particular wavelength back to the laser's active region. This stabilizes the output in a narrow range of wavelengths. Using a laser of this type, AT&T Bell Laboratories has demonstrated the transmission of 4 Gbit/s data through 103 km of single-mode fiber without the use of repeaters.

The complex structures needed to limit the emission linewidth (spectral width) pose problems in manufacture. The alternative is to reduce fiber dispersion, either by shifting the zero-dispersion wavelength to 1.55 μm or by spreading the dispersion

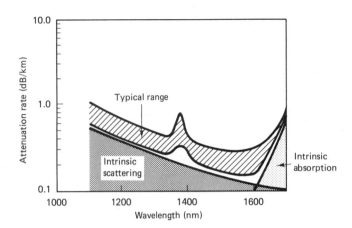

Figure 1.4 Spectral attenuation on a single-mode glass fiber.

minimum of the fiber out over a broader range. Refer to Figure 1.4. Note that the intrinsic scattering limit makes impossible an attenuation at 1.3 μm as low as at 1.55 μm. The basic material dispersion is difficult to alter. It is possible, however, to alter the waveguide dispersion by moving from a simple step-index design to those of a more complicated profile, as shown in Figure 1.3.

Strictly speaking, single-mode fibers are not single-moded. In fact, the two lowest-ordered modes with orthogonal polarization are supported. These two modes are degenerate in fibers that are circularly symmetric about their axis. That is, light in the two modes travels at the same speed, and light can couple (shift) between the two modes. Work is in progress on fibers that can discriminate between these polarization modes.

1.1.2.2 Material and waveguide dispersion. Material dispersion and waveguide dispersion are subtly different effects that arise because the manner in which light propagates through materials is a function of wavelength. Both types of dispersion are measured in picoseconds (of pulse spreading) per nanometer (of source spectral width) per kilometer of fiber length. These units reflect both the increase in magnitude with source linewidth and the accumulation of dispersion with distance in a fiber. The total dispersion, then, for a single-mode fiber is the sum of the two (see Figure 1.5).

Material dispersion is caused by the variation in refractive index of glass with wavelength. This is the same effect that causes a prism to spread out a spectrum. For this reason, it is sometimes referred to as *spectral* or *chromatic dispersion*. Material dispersion leads to pulse spreading even when different wavelengths follow the same path [see Equation 1.1].

Waveguide dispersion occurs in a single-mode fiber because light is not completly confined to the core. Single-mode fibers are discussed in Section 1.1.3. Roughly 20 percent of the light travels in the cladding adjacent to a step-index fiber. Since the refractive index of the cladding is less than that of the core, light in the core wants to travel faster. However, the core and cladding light belong to the same mode and thus must travel at the same speed. The effective velocity is somewhere

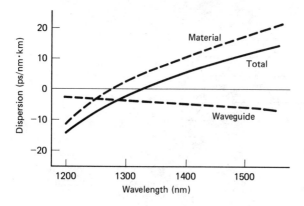

Figure 1.5 The total dispersion of single-mode fibers is a combination of material and waveguide dispersion. This produces a zero-dispersion point near 1330 nm where extremely large bandwidth is possible.

between the two speeds. Now, the effective velocity is wavelength-dependent; thus dispersion occurs. Note from Figure 1.5 that the change in waveguide dispersion with wavelength is smaller than that of material dispersion.

Material dispersion and waveguide dispersion can have different signs, which can result in a complete cancellation. In conventional germanium-doped silica fibers, the "zero-dispersion" wavelength is near 1.3 μm. The bandwidth of single-mode fibers becomes extremely large at this wavelength. This same effect is also important for graded-index fibers (see Section 1.1.4). Practically, the increase in dispersion with an increase in source spectral bandwidth is removed. Thus, we can use spectrally broadband LEDs at 1.3 μm without the high dispersion that their linewidth would cause at other wavelengths. At present, fiber makers are researching ways to combine the effects of low dispersion at 1.3 μm with the low attenuation at 1.5 μm, to produce new fiber types (see Figure 1.3).

1.1.3 Graded-Index Fiber

Graded-index multimode fibers were developed in the early 1970s as a compromise between the high bandwidth of single-mode fibers and the easy coupling of step-index multimode fibers. The core in graded-index fibers is radially graded in composition in contrast to step-index fibers. This results in the path length differences between refracting light rays to be minimized (see Figure 1.6). This effect greatly reduces multimode dispersion with a resultant increase in bandwidth. Transmission bandwidths greater than one gigahertz-kilometer are possible for 50-μm core fibers.

Graded-index multimode fibers with 50-μm cores and 125-μm cladding are currently used in many telecommunications systems operating at 45–140 megabits per second over distances in the 10–20-km range. For transmission at higher data rates, single-mode fibers are the preferred medium. A new graded-index fiber with an 85-μm core is available for short-distance video transmission and high-speed digital local-area networks. This fiber reduces the coupling problem, relative to the 50-μm core used for telecommunications, without a significant reduction in bandwidth.

In terms of minimum modal dispersion, analytical studies show that the optimum profile for a graded-index fiber is nearly parabolic. This is indicated in Figure

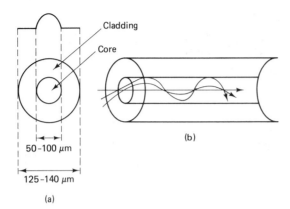

Figure 1.6 (a) Graded index fiber refractive index profile. (b) Light propagation in a multimode graded index fiber.

1.7, and most available graded-index fibers have this profile. Note that the refractive index tapers off parabolically from a value of n_1 on the central axis to n_2 at the outer radius. Modal dispersion for graded-index fibers can be as low as 2 ns/km. In terms of economy, graded-index fibers are between multimode and single-mode step-index fibers.

1.1.4 Impulse Response

The measurement of bandwidth based on the response of a system to an impulse is firmly embedded in linear systems theory. In terms of fiber optics systems, the impulse response is one measure of the dispersion within an optical fiber. This technique is illustrated in Figure 1.8. Narrow light pulses (impulses) are coupled into the

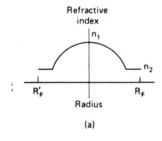

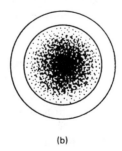

Figure 1.7 Graded index profile tapers off parabolically with radius from its central-axis value n_1 to a lower value n_2 at the fiber radius R_F. (b) Cross section shows that light travels more slowly near the shaded center region than it does away from the center, resulting in more consistent arrival time and less dispersion.

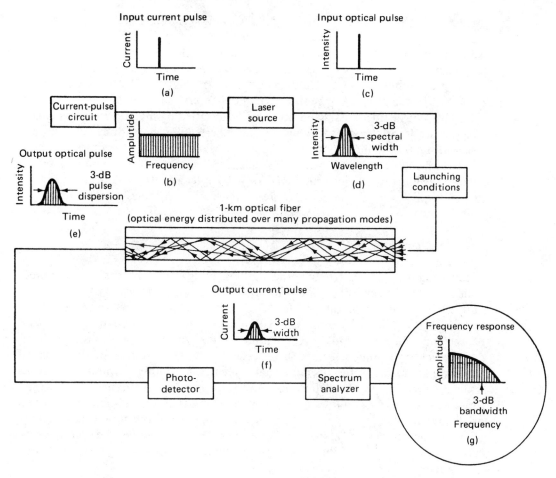

Figure 1.8 Inpulse response measurements determine the bandwidth of an optical fiber. (a) Narrow time-domain electrical pulses with (b) a flat frequency spectrum are converted to (c) light pulses with (d) a wavelength distribution characteristic of the laser source. Depending on the input conditions, light energy propagates through many modes of the fiber under test. Because of modal and material dispersion, (e) the output light pulse spreads with time and is photodetected as (f) a widened current pulse. (g) Frequency domain analysis then yields the fiber's 3-dB bandwidth.

fiber, and the output pulse width spread is measured as a function of the fiber length. The spread pulse is applied to a spectrum analyzer and the -3-dB bandwidth is measured. Theoretical studies show that the impulse response width resulting from modal dispersion is a function of the difference between the core and cladding indices. The modal impulse response for a step-index fiber varies linearly with this distance. For a graded-index fiber, the width is proportional to the square of the index difference. Since this difference is quite small, $n_1 - n_2 \cong 0.02$, a graded-index fiber achieves a considerable reduction in pulse spreading.

Dispersion is specified in nanoseconds/kilometer. This implies that the pulse spreading is proportional to fiber length. Actually, for fibers that are sufficiently long, the dispersion is proportional to the square root of the length. This incongruity has been attributed to an equilibrium-stated energy exchange (coupling) between the propagating modes. For fiber lengths L less than the coupling length L_c the dispersion is proportional to $\sqrt{LL_c}$. Unfortunately, while L_c is generally greater than one kilometer, it is usually not specified on manufacturers' data sheets.

The optical fiber's bandwidth is usually specified by manufacturers as either a bandwidth-length product with units of megahertz-kilometers, or as pulse dispersion with units of nanoseconds per kilometer. Manufacturers sometimes supply a complete frequency response curve, as illustrated in Figure 1.9. These two fibers have bandwidths of 200 and 400 MHz for a 1-km length. The -3-dB bandwidths for a 5-km link are 40 and 80 MHz, respectively. This assumes a monochromatic source.

1.1.5 System Losses

The starting point for determining the total loss budget for a system is the required optical power needed at the photodetector to ensure an appropriate signal-to-noise ratio (SNR) or bit error rate (BER). The SNR is applicable for analog systems, and the BER is commonly associated with digital systems. Because of optical losses between the transmitter and the receiver, only a fraction of the source's total radiant power reaches the photodetector. There are several principal sources of loss in signal power which result in the overall signal attenuation. Within a continuous fiber, structural and chemical imperfections cause absorption and scattering. Also, curvature of the fiber routes allows a degree of radiation that loses some of the modes originally launched. For short distances, this loss is insignificant and is generally ignored when attenuation is specified for a fiber. The remainder of the optical losses can be divided into the following: input-coupling losses, connector/splice losses, fiber attenuation, and output-coupling losses.

Input-coupling losses occur at the source/fiber interface. Usually, a short

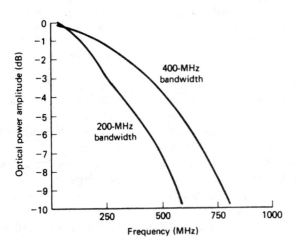

Figure 1.9 Fiber's 3-dB frequency is identified by taking the Fourier transform of the output signal. Specifications are usually quoted for a 1-km length.

length of fiber (called a *pigtail*) is permanently attached to the source's emitting area in single-fiber communications. Any mismatch that exists between this emitting area and the pigtail's core area results in the first input-coupling loss. Additional loss will occur if the core area is smaller than the source emitting area. The fractional coupling loss is approximately the ratio of the core area to the emitting area. A second input-coupling loss factor relates to the light-gathering ability of the fiber itself. The *numerical aperture* (NA) defines the acceptance-cone angle θ of the fiber (see Figure 1.10). Mathematically, the numerical aperture relation acceptance cone may be expressed as

$$\text{NA} = \sin \theta \tag{1.5}$$

Example 1.3

Find the numerical aperture and acceptance-cone half-angle for the fiber of Example 1.1 (see Figure 1.11).

Solution From Snell's law:

$$n_1 \sin \theta_1 = n_2 \sin \theta_2$$

Thus

$$n_1 \sin \theta_{\min} = n_2 \sin 90°$$

and

$$\sin \theta_{\min} = \frac{n_2}{n_1}$$

By definition, the numerical aperture is

$$\text{NA} = \sin \theta_a$$

Thus, from Snell's law:

$$n_0 \sin \theta_a = n_1 \sin (90° - \theta_{\min}) = n_1 \cos \theta_{\min}$$

$$= n_1 \sqrt{1 - \sin^2 \theta_{\min}} = n_1 \sqrt{1 - \left(\frac{n_2}{n_1}\right)^2}$$

For air, $n_0 = 1$:

$$\text{NA} = \sin \theta_a = \sqrt{n_1^2 + n_2^2} = \sqrt{1.48^2 - 1.46^2} = 0.242$$

Acceptance-cone half-angle $= \sin^{-1} 0.242 = 14.03°$

In order to evaluate the NA input-coupling loss, a knowledge of the source's emission profile is needed (see Section 1.2.1). An additional input-coupling loss referred to as the *packing fraction* occurs in working with a fiber bundle (many fibers grouped together and illuminated by one light source). This loss is measured by the ratio of the collection-core areas to the total bundle cross-sectional area. The packing fraction loss in decibels is ten times the log of this ratio.

As discussed previously, single-fiber development is progressing so rapidly that

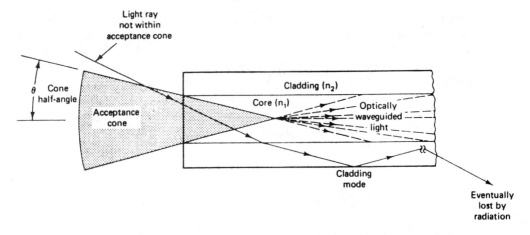

Figure 1.10 Fiber's acceptance-cone half-angle derives from the fiber's numerical aperture. Light projected into the acceptance cone undergoes waveguiding in the core, while rays outside the cone reflect into the cladding and are eventually lost.

this is likely to be the choice for data transmission. Even so, bundles are the most efficient for both collecting and distributing light in either a random or formatted mode. The bundle's greater energy is an advantage for applications such as scanners, card readers, and medical instruments. It appears that ultimately fiber optic bundles will become a custom business.

The last and least important input-coupling loss results from light reflection at the input end of the fiber. This loss is on the order of 0.2 dB. The output-coupling losses are generally not as severe as the input-coupling losses. These losses are generally on the order of 1 dB.

Between the system input and output occur fiber attenuation losses as well as those arising from splices and connectors. In order to minimize the fiber loss, we desire a source whose wavelength falls in the area where fiber attenuation is low. We noted previously that this is in the area of 1.3 to 1.5 μm (see Section 1.1.2.1). An attenuation-vs-wavelength profile for two of Corning's fibers is shown in Figure 1.12. Fortunately, the emission and detection characteristics of light-emitting diodes (LEDs), laser diodes, and various silicon detectors are quite compatible with the transmission characteristics of common fiber core materials.

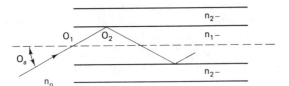

Figure 1.11

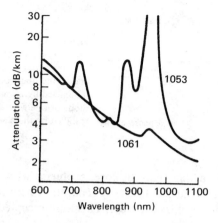

Figure 1.12 Fiber attenuation varies dramatically with wavelength, so good overall system performance requires a good match between source and fiber characteristics.

1.2 DETECTORS AND SOURCES

In the preceding sections, we concentrated on the parameters of the optical transmission media. This discussion provides an understanding of basic communication link limitations over fiber optic links with respect to optical power loss and dispersion. To complete this discussion we also need an understanding of optical sources and detectors. Before considering the light source, it is instructive to consider first the parameters that affect the input efficiency and also to learn how to calculate the losses associated with this parameter.

Next to fiber attenuation, input coupling contributes most to system loss. We can categorize the losses as:

- Unintercepted-illumination (UI) loss resulting from an area mismatch between the source's illumination spot (in the plane of the fiber end) and the fiber core area
- Numerical-aperture (NA) loss arising from light rays with angles of incidence not within the fiber's acceptance cone
- Reflection (R) loss from the end of the fiber

In general, the magnitude of the input-coupling loss depends on the geometrical and optical characteristics of the source and fiber. Parameters that are particularly important are the source emitting area and angular emission profile, the fiber core area, the refractive index and numerical aperture, and the distance between the emitting surface and the fiber end.

Area mismatch between the source's emitting area and the fiber can result in unintercepted illumination. In particular, if the source's emitting area is larger than the fiber core area, all of the light cannot be coupled into the waveguide. Therefore, we wish to choose a source whose emitter is no larger than the core. Fortunately, there are available both LEDs and ILDs (injection laser diodes) that fit this require-

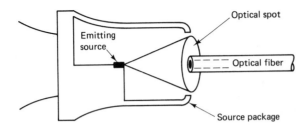

Figure 1.13 Unintercepted illumination loss can be a problem when the light-emitting surface is separated from the end of the fiber core. The loss can be minimized by using an uncapped source diode or one with a "pigtail" already installed by the manufacturer.

ment. Even if the source is smaller than the core, problems with unintercepted illumination may still exist (see Figure 1.13). Any separation between the source and the fiber allows emitted light to miss the fiber. This unintercepted loss (UI) is given approximately as

$$UI_{loss} = 10 \log \frac{A_c}{A_s} \tag{1.6}$$

where

A_c = fiber core area

A_s = area of the source's projected optical spot in the plane of the fiber end

The magnitude of UI loss depends on several factors. Specifically, these are the source's angular emission profile, the distance between the source's emitting surface and the fiber end, and the diameter of the core. All small sources have rapidly divergent beams. Thus, in order to avoid intolerable losses, separation between the source and the fiber end can be no greater than two to four times the core diameter. This requirement effectively eliminates using a glass-capped diode, since the source pellet typically lies 1 or 2 mm in back of the window. The recourse, then, is either to buy an uncapped diode and then mount the fiber as closely as possible to the emitting surface or to obtain a source with the pigtail already installed by the manufacturer (see Figure 1.13).

1.2.1 Source Launch Profile

With the UI losses overcome, we are in position to consider losses associated with the launch profile. Specifically, this is reduced performance arising from the numerical aperture [see Equation (1.5)]. Refer to Figure 1.14. Note that considerable light is lost because of the conflict between the relatively small acceptance cone angle of a fiber (half-angle on the order of 10 deg to 14 deg) and the broad divergence of both LED and ILD emission beams. In order to estimate the input-coupling efficiency (NA loss), it is necessary to first describe the source-beam profile. Unfortunately, this information is not always supplied by manufacturers in the form needed. Often it can be deduced from the specifications.

Typically, manufacturers plot the source-beam profile on a polar diagram.

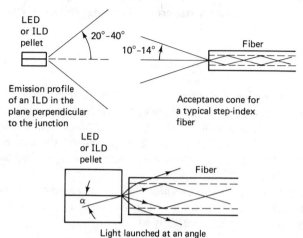

Figure 1.14 Lots of light is not necessarily a blessing at the source-fiber interface. Both LEDs and ILDs have broadly divergent emission beams, and all light radiated outside of the fiber's acceptance-cone angle contributes to numerical aperture loss.

The curves are derived by measuring the relative radiant intensity on a small detector that swings through a 180-deg arc. As an example, the intensity of a uniform source emitter varies with the cosine of the angle between a line perpendicular to it and another line to the observation point [Figure 1.15(a)]. This emission power profile is produced by a *Lambertian source*. We can describe the emission power profile of a Lambertian source as

$$P = P_0 \cos \phi \tag{1.7}$$

where P_0 is the radiant intensity along the line $\phi = 0$.

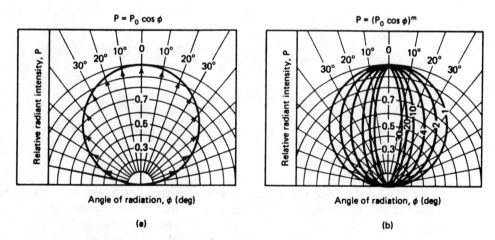

Figure 1.15 A wide variety of emission-beam patterns appear in sources available today. (a) While uniform surface emitters typically have Lambertian-type profiles, (b) some sources exhibit much narrower beam profiles.

Sec. 1.2 Detectors and Sources

Other sources produce narrower beam patterns. Mathematically, we can approximate these patterns as

$$P = P_0(\cos \phi)^m \tag{1.8}$$

and a set of curves, plotted for various values of m, is illustrated in Figure 1.15(b). Using the manufacturer's data sheet, we can use these curves to find an approximate value for m. In some cases, manufacturers may specify only the 50-percent intensity points (half-angle beam spread). Interpolation can be employed on the curves of Figure 1.15(b). For example, if the 50-percent intensity points were specified as 25 deg, the value of m is approximately 7.

The question naturally arises as to the usefulness of the parameter, m. With this quantity we are able to calculate the amount of source power P_C, coupled into a fiber. We can express this relation mathematically as

$$P_C = P_T[1 - (\cos \theta)^{m+1}] \tag{1.9}$$

where

$$P_T = \text{source power, mW}$$

$$\theta = \text{acceptance-cone half-angle}$$

Occasionally, specification sheets for sources relate the cone half-angle to the percentage of total radiant flux within a fiber cone (see Figure 1.16). For this particular example, a fiber with an acceptance cone of 14 deg (NA = 0.25) would capture about 28 percent of the total radiated power, P_T.

Example 1.4

Calculate the aperture loss for the fiber in Example 1.1 if a Lambertian source is used.

Solution

$$P_T = P_C[1 - (\cos \theta)^{m+1}] = P_T[1 - \cos^2 \theta]$$

$$= P_T \sin^2 \theta = P_T(\text{NA})^2$$

$$\frac{P_C}{P_T} = (\text{NA})^2 = (0.242)^2$$

$$\text{NA}_{\text{loss}} = 10 \log \frac{P_C}{P_T} = 10 \log (0.242)^2 = 12.32 \text{ dB}$$

1.2.2 Reflection Loss

We noted earlier that reflection loss occurs at the ends of the fiber. Compared to the NA loss, the reflection loss is almost negligible. It is important, however, in fiber splices. Light incident on the fiber core experiences a change in the index of refraction at the air/core interface. Thus, part of the light reflects back from the end surface and is lost. Reflected/refracted proportioning of the incident rays depends

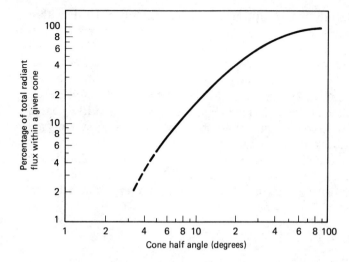

Figure 1.16 Specification sheets can be useful, as evidenced by this curve occasionally found on source data sheets. By relating cone half-angle to the percentage of total radiant flux within a given cone, designers can calculate the amount of light captured at the source-fiber interface.

on the core index of refraction. We can define a *reflection coefficient,* ρ, which gives the fraction of incident light reflected from the core. This relation is approximately

$$\rho = \left(\frac{n_1 - 1}{n_1 + 1}\right)^2 \quad (1.10)$$

Derivation of this equation is based on the classical Fresnel formulas for normal incidence. In terms of decibels, the reflection loss is given by

$$R = 10 \log (1 - \rho) \quad (1.11)$$

Example 1.5

Calculate the reflection loss for the fiber of Example 1.1.

Solution

$$\rho = \left(\frac{n_1 - 1}{n_1 + 1}\right)^2 = \left(\frac{1.48 - 1}{1.48 + 1}\right)^2 = 0.0374 \quad (3.74\%)$$

$$R = 10 \log (1 - \rho) = 10 \log (1 - 0.0374)$$

$$= -0.166 \text{ dB}$$

1.2.3 Source Requirements

Requirements are severe for optical sources that must function with long-haul single-fiber transmission media. In order to maximize the link length between repeater spacing, a source should be as intense as the state of the art allows. In addition, as noted previously, the source emitting area should be smaller than the fiber core area in order to efficiently launch optical power into the fiber. To fit within the fiber's acceptance cone, the emitted beam pattern should be very directional. In fact, it should be almost collimated and nearly monochromatic in order to avoid material dispersion. Modern high-capacity digital systems require rise and fall times in the nanosecond range. In terms of analog systems, the optical output power should be linearly related to the drive current or voltage over a wide dynamic range.

Semiconductor injection lasers (ILDs) and light emitting diodes (LEDs) based on the ternary material (GaAlAs) and quaternary material (GaLnAsP) systems are used as optical sources in fiber optic communication systems. The power output for surface-emitting devices is usually specified in terms of "radiance." This is the power per unit of solid angle (steradian) per unit area (W/sr-cm^2). Example 1.6 illustrates a typical system calculation for calculating total power radiated by the device.

Example 1.6

The radiance of an LED is specified to be 60 W/sr-cm^2. For Example 1.1, assuming a Lambertian source, calculate: (a) total power emitted into a solid angle 2π steradians, (b) P_C, and (c) NA$_{loss}$.

Solution

(a) $P_T = (60 \text{ W/sr-cm}^2)(2\pi \text{ sr})(\pi R^2)$

$\quad\ = (60)(2\pi)(\pi)(25 \times 10^{-4})^2 = 7.4$ mW

(b) $P_C = P_T[1 - (\cos \theta)^2]$ mW

From Example 1.3, $\theta = 14.03$ deg.

$$P_C = 7.4[1 - (\cos 14.03°)^2] = 0.43 \text{ mW}$$

(c) $10 \log \dfrac{P_C}{P_T} = 10 \log \dfrac{0.43}{7.4} = -12.3$ dB

The peak emission wavelength is another important source characteristic. Ideally, this should match the fiber's minimum attenuation wavelength. Sources constructed from GaAlAs radiate in the wavelength range of 800 to 900 nm, while sources made from GaLaAsP emit in the range from slightly above 1000 nm to 1700 nm. The exact emission wavelength depends on the device's material composition. From previous discussion, we see the importance of the latter devices. As previously noted, a minimum occurs in the chromatic dispersion of fibers at 1300 nm, and fiber absorption loss is a minimum at 1550 nm (less than 0.2 dB/km).

For moderate distance data links, an LED source is generally appropriate. There are currently three popular types of light emitting sources:

- Burrus-type
- Edge emitter
- Surface emitter

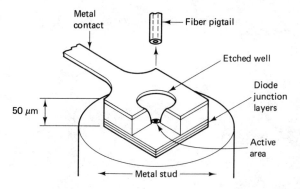

Figure 1.17 Fiber-source interfacing occurs in the etched well in this Burrus-configured surface emitter. Fiber pigtail, placed in the well in intimate contact with the emitting area, is epoxied into place.

In 1971, C. A. Burrus of Bell Laboratories configured a high-radiance LED with a small circular surface (Figure 1.17), and an etched well on one side to allow attachment of a single fiber. These surface emitters approximate a Lambertian source, suffering high NA loss as a result. An edge- or side-emitting diode is illustrated in Figure 1.18. In this configuration, guiding layers channel the light toward the fiber core to produce a narrower beam source with significantly reduced NA losses. Junction thickness is typically about 1 or 2 μm. A stripe contact restricts the width of the active area to 10 to 20 μm. This process results in a rectangular active source area smaller than the fiber core. Theoretical studies indicate that edge-emitting diodes should be able to output three to seven times the power of a surface emitter. Surface emitters also radiate from the edge similarly to the illustration in Figure 1.18.

Light emitting diodes are less efficient and more nonlinear than lasers. In addi-

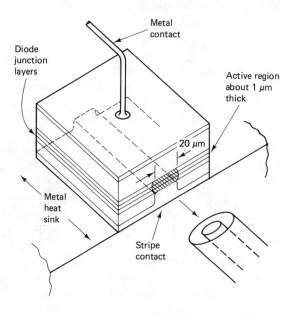

Figure 1.18 Easy fiber-source coupling occurs in stripe contact edge-emitting sources. However, some broad-area (no stripe) edge emitters suffer from unintercepted-illumination loss.

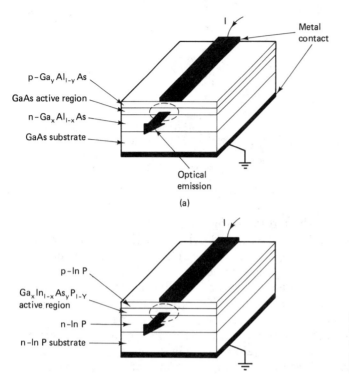

Figure 1.19 Schematic diagrams of (a) GaAlAs semiconductor injection laser and (b) InGaAsP laser. (Reprinted with permission of *Microwave Journal,* April 1985, © Horizon House-Microwave, Inc.)

tion, LEDs exhibit a broad emission spectrum and typically have a direct modulation bandwidth less than 100 MHz. Thus, LEDs are not of interest in high-speed microwave applications. For high speeds, ILDs are the preferred choice. Schematic diagrams for laser diodes utilizing the two types of material are shown in Figure 1.19.

The active region is a very thin (~ 0.2 μm) semiconductor layer where coherent light is generated. The neighboring crystal layers that sandwich the active layer are higher bandgap materials for confining carriers and photons in the active layer of the laser. This maximizes the interaction cross section of the carriers and photons and improves device efficiency.

In a semiconductor laser, the mirrors that form the optical cavity are constructed by cleaving two parallel facets of the semiconductor crystal. Photons are generated within the cavity by spontaneous recombination of electron–hole pairs. These photons circulate within the laser cavity, thus stimulating the emission of additional coherent photons, that is, photons that have the same wavelength and phase. As the injection current, i, increases, the optical gain eventually overcomes optical losses in the resonator and the device becomes an oscillator. Under these conditions, emission with a narrow spectral width is obtained. The minimum current at which this phenomenon occurs is called the *lasing threshold current,* I_{th}. A typical optical power versus injection characteristic for a semiconductor is shown in Figure 1.20. Also shown in this figure is a typical current–voltage characteristic. Note that,

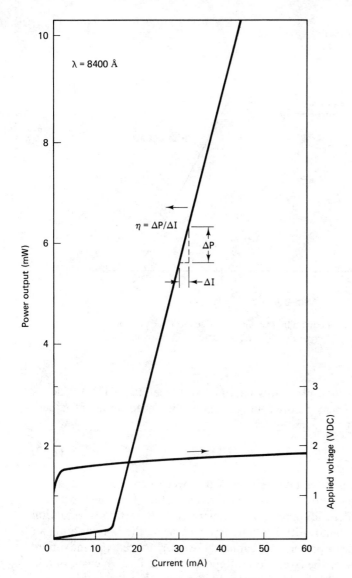

Figure 1.20 Optical power output versus current and current–voltage characteristics of a laser diode. (Reprinted with permission of *Microwave Journal,* April 1985, © Horizon House-Microwave, Inc.)

above the lasing current threshold, the scope of the optical output power versus injection current curve represents the quantum efficiency of the device. That is,

$$\eta = \frac{\Delta P}{\Delta i} \text{ W/A} \qquad (1.12)$$

where η = quantum efficiency (number of photons generated per unit number of electrons injected). High-quality lasers are characterized by low threshhold currents and high quantum efficiencies.

Theoretical analysis has shown that the intrinsic modulation response of a

Sec. 1.2 Detectors and Sources   23

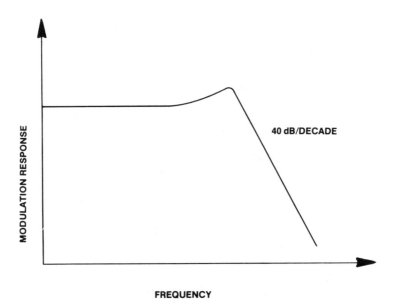

Figure 1.21 Theoretical modulation response of a semiconductor diode. (Reprinted with permission of *Microwave Journal,* April 1985, © Horizon House-Microwave, Inc.)

laser diode behaves as a second-order low-pass network (see Figure 1.21). The response exhibits a resonance peak before rolling off at 40 dB/decade at high frequencies. The -3-dB bandwidth is given by

$$f_{-3dB} = A\sqrt{P_{out}} \text{ GHz} \tag{1.13}$$

where P_{out} is the continuous wave (average) optical power of the laser and A is a parameter that depends on the structure of the laser. The value of A ranges between about 1 and 4 GHz/mW, depending on device construction. Referring to Equation (1.13), we see that a larger bandwidth is obtainable by biasing the laser at a higher continuous-wave (cw) optical power. We cannot, however, exceed the maximum rated optical output power of the laser diode. Refer to Figure 1.22. We can obtain a higher modulation bandwidth by biasing at a higher level, but the amplitude of the modulation signal applied to the laser is reduced correspondingly. Depending on the bandwidth of the particular application involved, we can choose an optimum bias point where the modulation bandwidth of the laser is adequate and where the modulation signal can be optimized.

The laser diode must also be matched to the system driving it. A forward-biased laser diode has a very low impedance—basically resistive, of a few ohms. If we wish to match to a 50-ohm system, a resistor slightly less that 50 ohms can be placed in series with the diode. At first glance, this would appear to destroy the efficiency of the device, since most of the drive power is delivered to the resistor and not to the laser diode. However, it should be noted that the optical emitted power is affected by the injected current, and not the power dissipation. Power

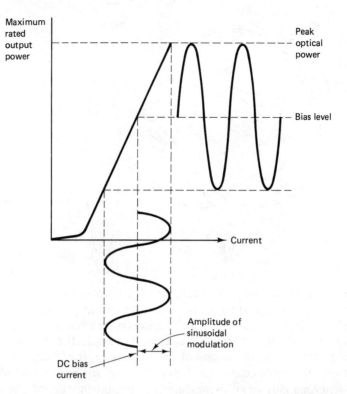

Figure 1.22 Modulation of the optical emission from a laser by modulating the input drive current. The maximum rated power should not be exceeded at any point. (Reprinted with permission of *Microwave Journal*, © Horizon House-Microwave, Inc.)

dissipation serves only to heat up the device and degrade its performance. Under matched conditions, the current delivered into the laser diode for a given drive power is approximately half that of an unmatched laser. This loss in drive efficiency can be justified if substantial problems caused by reflections from the device can be avoided.

Solid-state laser diodes are exceptionally well suited for use in digital fiber optic systems. Stripe contact versions have active emitter regions measuring about 1×20 μm. Emission through the slit produces a rather broad beam in the plane perpendicular to the junction (20-deg to 40-deg half-angle). The beam in the plane parallel to this junction is much more narrowly defined (Figure 1.23).

1.2.4 Detector Requirements

In fiber optic systems, the most commonly used receivers utilize photodiodes (either PIN or avalanche types) to convert incident light into electrical energy. A photodetector "demodulates" an optical signal by generating a current proportional to the intensity of the optical radiation, thereby converting the variations in optical intensity into an electrical signal. Before considering detector requirements, it is of inter-

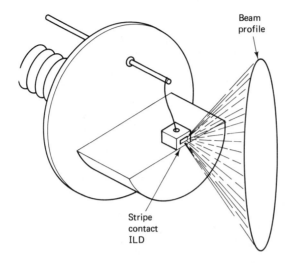

Figure 1.23 Significant NA-loss reduction results from the narrow beam profile (in the plane parallel to the junction) of stripe-contact-configured laser diodes.

est to look at diode characteristics. A PIN photodiode consists of a large intrinsic (very lightly doped) region sandwiched between p- and n-doped semiconducting regions. Photons absorbed in this region create electron–hole pairs that are then separated by an electric field, thus generating an electric current in the load circuit (see Figure 1.24). As in the case of the laser diode, the efficiency of the optical-photon-to-electron-hole conversion process is specified by the photodiode's *quantum efficiency*, η. This quantity measures the average number of electrons released by each incident photon. A number near one indicates a highly efficient diode. In general, the quantum efficiency is a function of wavelength and temperature. It is obvious that this quantity enters directly into the computation of the overall loss of a fiber-optic link.

The efficiency of a photodetector is generally expressed in terms of its *responsivity*, R_λ. This parameter is related to the quantum efficiency, η, by the following relation:

$$R_\lambda = \eta\lambda/1.24 \text{ A/W} \tag{1.14}$$

where λ is the light wavelength. In words, the responsivity is the ratio of the output current to the input optical power, measured in amperes/watt. The signal current,

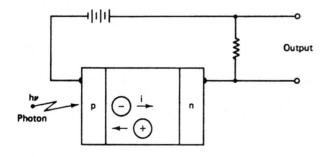

Figure 1.24 Electron–hole pairs, created by photons absorbed in the intrinsic region of a PIN diode, are separated by the diode's electric field to produce a current in the load circuit.

i_s, generated when P watts of optical power are incident on a photodiode with responsivity r is

$$i_s = R_\lambda P \text{ amperes} \qquad (1.15)$$

For typical PIN diodes, peak wavelength responsivities are less than 1 A/W.

An *avalanche photodiode* (APD) is designed for applications requiring greater sensitivity. An APD has the same basic structure as a PIN diode. When the reverse bias to the diode is increased, the electric field in the intrinsic (depletion) region increases correspondingly. When the electric field becomes sufficiently high (on the order of 10^5 V/cm), an electron or a hole can collide with a bound electron with sufficient energy to cause ionization, thereby creating an extra electron–hole pair. The additional carriers, in turn, can gain enough energy from the electric field to cause further impact ionization until an avalanche of carriers is produced. Thus, a single incident photon light can create G_0 electrons, and the responsivity of the avalanche photodetector is increased by the multiplication factor, G_0. Bias voltages for APDs are considerably higher than those required for PIN diodes. Bias voltages on the order of 300 V are not uncommon, versus a few volts for PIN photodiodes. Typical gain values for APDs are on the order of 100 to 150.

The materials used to construct both PIN and APD photodiodes reflect the wavelength range of interest. Silicon photodiodes can detect radiation from visible up to about 1 μm. Germanium photodiodes respond up to 1.6 μm, while GaAs photodiodes cover the range from 0.7 to 0.9 μm. A Schottky-barrier photodiode is also available. It works on the same general principle as a PIN diode, but optical absorption takes place at the depletion region of the metal-semiconductor Schottky junction instead of a separate intrinsic layer. The responsivity of these detectors is somewhat lower due to the presence of a metal layer on the surface of the photodiode. Responses up to 100 GHz have been reported for these diodes. PIN diodes are available with bandwidths up to 7 GHz. The bandwidth of avalanche photodiodes is dependent on carrier transit time in the high field avalanche region and on carrier ionization rates. It is intuitively apparent that collision of carriers in the avalanche process inevitably slows down the carrier speed and thus reduces the bandwidth of the detector. Theoretical studies have shown that for high values of G_0, the relation between G_0 and bandwidth follows approximately a constant gain–bandwidth product.

Generally speaking, in the design of fiber optic transmission systems, where very long-distance transmission (greater than 100 km) is involved, an avalanche photodiode is selected. A 15- to 20-dB advantage in sensitivity is available to compensate for the weak optical power. For shorter distances where the optical power is strong, a PIN diode is preferred.

In the receiver circuit, the signal current must contend with noise currents. The dominant noise component in PIN photodetectors is caused by fluctuations in "dark current." This is the current that flows through the diode biasing current when no light is incident on the photodiode. Usually manufacturers specify an average dc value for dark current at a given temperature and bias voltage. Theoretical studies have shown that dark current shot-noise power varies linearly with this average. It also increases with temperature. As a general rule, it doubles with every 10°C increase in operating temperature.

Another photodiode figure of merit related to noise performance is the *noise equivalent power,* NEP. The noise equivalent power is usually expressed in watts/$\sqrt{Hz}$. If we multiply NEP by the square root of the detector bandwidth, *B*, an absolute power is obtained called the *minimum detectable signal,* MDS. This quantity, MDS, defines the required optical power incident on the photodiode to generate a photocurrent equal to the total photodiode noise current, that is, a 0-dB signal-to-noise ratio at the output of the photodiode. Obviously, the receiver must be operated at a level higher than the MDS.

Because the current generated by the photodiode is so small, a pre-amplifier is needed following the photodetector. This amplifier is also a noise source and is not included in the photodiode's NEP. Data rates are limited by the response time. In general, the limiting factor is the RC time constant associated with the diode's resistance plus the load resistance into which it operates and the junction capacitance. Typically, PIN diode's 10 to 90 percent rise time is a few nanoseconds. This topic is discussed in Section 1.3.8.

1.2.4.1 Light reception. As light exists from the end of an optical fiber, it spreads out with a divergence that is approximately equal to the acceptance cone of the fiber. This is determined by the NA of the fiber. The photodiodes are packaged with their photosensitive surfaces located a distance *S* behind a protective glass window (Figure 1.25). Any light that expands beyond the active area of the photodiode represents a coupling loss.

Example 1.7

(a) A fiber has an exit angle $\theta = 14$ deg and a core diameter of 0.05 mm. It is to be coupled to a photodiode whose circular sensitive area is a 5 mm² and which is located 2.5 mm behind the glass window. Calculate the coupling loss.

(b) A second fiber (NA = 0.6) is used to gather more light at the input end. With all parameters the same as part (a), calculate the coupling loss.

Solution

(a) The light area at the photodiode surface is

$$A = \pi r^2 = \pi \left(S \tan \theta = \frac{d}{2}\right)^2$$

$$= \pi \left(2.5 \tan 14° + \frac{0.05}{2}\right)^2 = 1.33 \text{ mm}$$

This figure is smaller than the detector area. There is no coupling loss due to expansion of the light beam.

(b) $\qquad$ NA $= \sin^{-1} (0.6) = 37°$

$$A = \pi \left(2.5 \tan 27° + \frac{0.05}{2}\right) = 11.55 \text{ mm}^2$$

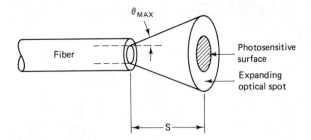

Figure 1.25 Unintercepted illumination is also a potential problem at the receiving end of a link. Using a photodiode with a large active area keeps this loss down. A better solution involves using an uncapped diode and moving the fiber as close as possible to it.

$$\text{Area ratio} = \frac{5}{11.5} = 0.435$$

$$\text{Coupling loss} = 10 \log 0.435 = 3.6 \text{ dB}$$

One way to reduce this loss would be to remove the diode's glass window and move the fiber closer to the detector. As an alternative, a photodiode with a larger active area could be selected.

1.3 SYSTEM DESIGN PARAMETERS

In the previous sections, we considered the characteristics of individual components such as fibers, sources, and detectors. It is instructive at this point to examine some general system issues. Among these are tradeoffs in such factors as component selection, component compatibility, system configuration, multiplexing methods, types of modulation, and system performance.

1.3.1 Component Selection

An evaluation of individual components is necessary in any system design. The choice of an optical source—LED or ILD—is the first area in which a tradeoff must be made. Because an ILD radiates more power and has a narrower beam profile, it allows longer optical links without repeater spacing. In addition, for all practical purposes, the narrow spectral width allows us to ignore fiber-material dispersion. An ILD's faster response time permits higher modulation rates than an LED. The disadvantage, as mentioned previously, is a higher biasing voltage.

The next consideration is the available types of fiber. The primary U.S. fiber standards organization has identified the fiber classes shown in Table 1.1. Class I all-glass fiber is by far the most common variety, with attenuation coefficients, depending on wavelength, from below 1 to about 4 dB/km and bandwidths exceeding 1 GHz-km for the graded Class Ia. Class II fiber with a glass core and plastic clad is capable of about 6 dB/km and bandwidths up to 40 MHz-km. These values are not as good as those for Class I, but the fibers possess greater strength. Class III all-plastic fiber is least common, with attenuation exceeding 100 dB/km at selected wavelengths. Class IV offers the advantage of extremely high bandwidths and low attenuation. Its disadvantage is poor numerical aperture because of its size.

TABLE 1.1 ELECTRONIC INDUSTRIES ASSOCIATION STANDARD RS-458-A OPTIONAL WAVEGUIDE FIBER MATERIAL CLASS STANDARDS

General Type	Class	Index	Core Material	Cladding Material	Jacket Material
Multimode	Ia	Graded	Glass	Glass	—
	Ib	Quasi-step	Glass	Glass	—
	Ic	Step	Glass	Glass	—
	IIa	Step	Glass	Plastic	—
	IIb	Step	Glass	Plastic	Plastic
	III	Step	Plastic	Plastic	—
Single mode	IV	To be determined	Glass	Glass	—

In terms of the receiver, we can choose between a PIN diode and an avalanche photodiode (APD). Because of its high responsivity, the APD is the choice where high signal bandwidth and low power detection are critical. In choosing components for a fiber link, it is important to ensure that the choices are wavelength compatible. Optimum system performance demands that the source's emission-peak wavelength match the valleys in the fiber's attenuation–wavelength characteristic. These characteristics for multimode and single mode are shown in Figure 1.26. In addition, the detector must be responsive in this range. Fortunately, devices that meet these requirements are available as off-the-shelf components.

1.3.2 System Architecture

For a fiber optic system, the simplist architecture is a point-to-point configuration with an access point at each end. Bidirectional communications may require the use of multiplexing or a two-cable layout. This simple architecture, however, does not

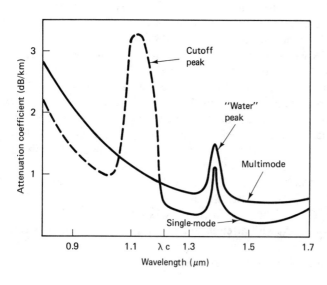

Figure 1.26 Spectral attentuation coefficients for multimode and single-mode fiber.

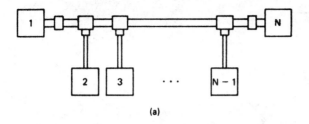

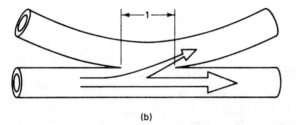

Figure 1.27 (a) Terminals tap off or inject optical energy onto the main trunk in the in-line bus configuration. (b) One approach to the T-coupler implementation of this configuration is to fuse the fibers together; the amount of light coupled varies with interaction length, l, and core-to-core proximity.

suit some applications. Today systems may require communication between several geographically distributed terminals. In this case, a multiple-access configuration is in order. An additional fiber optic component called a *coupler* is needed for this type of architecture. There are two coupler options—the T and the star. These devices are used extensively in local area networks (LANs). The T-coupler design is illustrated in Figure 1.27. Optical energy can be tapped off or injected onto the main trunk at each coupler. A star configuration is illustrated in Figure 1.28. The operation is as follows: A subscriber transmits a signal on a dedicated fiber toward the coupler. The light enters the coupler at port 1, spreads out and is reflected by the dielectric mirror into all the remaining ports.

In terms of a comparison, the optical signal suffers a loss at each T-coupler in the bus. Worst-case system loss (in decibels) increases linearly with the number of terminals. On the other hand, the star configuration introduces only one coupler loss. Furthermore, as the number of ports, n, increases, the star coupler's power-splitting loss increases only as $10 \log n$. When T-couplers with a constant tap loss are used, dynamic range problems can arise. For example, the transmitter must output enough power to drive the most remote station while simultaneously not overdriving the nearest.

In the majority of star-coupler applications, the two important measures of a coupler are total loss and the leg-to-leg uniformity of the light output. Coupler loss is quoted in one of the two ways: excess loss (EL) or insertion loss (IL). Excess loss is defined as

$$EL = -10 \log \left[\frac{\Sigma P_i}{P_o}\right] \quad (1.16)$$

where P_o is the light intensity launched into the coupler and P_i is the intensity out of a given fiber leg i. The intensities are summed over each of the output fibers. The insertion loss for a single leg is defined as

Sec. 1.3 System Design Parameters

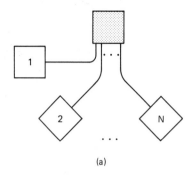

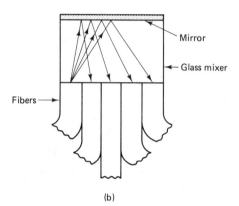

Figure 1.28 One terminal can communicate with all others when you employ the star-coupler configuration. (b) Light entering at one port spreads out in the glass mixer and reflects from the mirror into all other ports.

$$\text{IL} = -10 \log \frac{P_i}{P_o} \qquad (1.17)$$

The variation in insertion loss defines the nonuniformity of the coupler.

A loss comparison for the two types of couplers is shown in Figure 1.29. In this figure, worst-case system losses are plotted against the number of terminals for both the T (in-line) and star coupler. A constant 10-dB tap loss is assumed plus a 2-dB insertion loss for the T-coupler. A 7-dB insertion loss is assumed for the star design. Values of 1 and 3 dB are assumed for connector and I/O splitting losses, respectively. From this figure, we see that the star design has lower loss with more than about four terminals.

1.3.3 Signal Multiplexing

If several different signals are to be transmitted, some form of multiplexing is required. One possibility is *space-division-multiplexing* (SDM). In this technique, a separate fiber is used for each signal. This technique works fine for a small number of signals. Frequency- and time-division-multiplexing (FDM and TDM) are familiar schemes that can be used to combine several channels onto one fiber. In FDM, several channels are interleaved in the RF (rather than optical) domain, and the

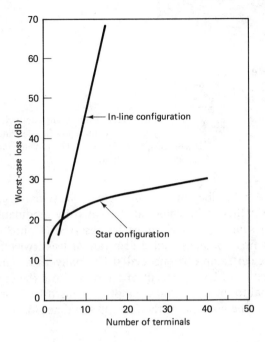

Figure 1.29 The star configuration proves superior as the number of terminals increases. Because there is only one coupler loss to contend with in this configuration, this result is not too surprising.

composite RF signal intensity modulates a light source. In TDM, signals are interleaved over time. A newer technique, called *wavelength-division-multiplexing* (WDM), uses sources of distinct wavelengths to separate various channels. Optical filters are required at the photodetectors to select the appropriate channel. This technique further increases the bandwidth advantage of optical fibers.

1.3.4 Modulation

We can modulate both an LED and ILD simply by varying the current flowing through them. Several methods are available for accomplishing this. In terms of analog information, the most obvious option is *baseband analog-intensity modulation*. In this technique, an input voltage signal (voice or video) at baseband is converted to current variations through an LED or ILD. The output optical power varies in almost direct proportion with the electrical input signal. The diode must be biased at the appropriate level to avoid overmodulation (clipping). Signal detection occurs through a square-law (optical power-to-electrical current) photodiode (PIN or APD) and a current-to-voltage front end amplifier.

In some designs, the electrical input interface may not be at baseband. An example is subcarrier analog-intensity modulation (Figure 1.30). In this particular scheme, the baseband analog signal modulates an RF subcarrier, which then intensity-modulates the optical source. Obviously, any of the familiar schemes AM, SSB, or FM could be used to modulate the intermediate subcarrier. Appropriate modulation indices must be selected to prevent overmodulation. For analog signals of reasonable bandwidth, we may consider *pulse-position-modulation* (PPM). Analysis shows that

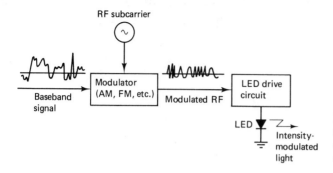

Figure 1.30 Using subcarrier analog modulation, you first modulate an RF carrier and then modulate the LED output.

by exploiting the bandwidth of fiber optics, an increase in signal-to-noise ratio (SNR) can be achieved over that of baseband analog-intensity modulation. When PPM is used, the Nyquist sampling criteria must be observed (Figure 1.31). The baseband analog signal is first voltage-sampled at a rate at least twice the analog signal bandwidth. Next, each sample is encoded positionally as a narrow pulse within a time slot dedicated to it. Displacement of the pulse from the center of the nth time slot is proportional to the nth voltage sample. A pulsed LED or ILD then transmits this information. The position of the pulse in the time slot now contains information.

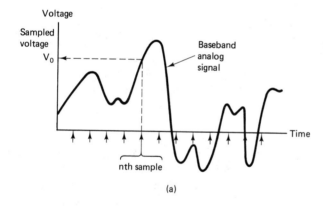

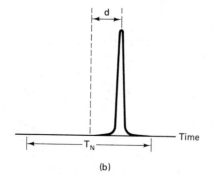

Figure 1.31 You can exploit the available bandwidth of fiber optics by using pulse-position modulation. (a) First sample the analog signal, and then (b) encode the nth sample as a pulse of light during the time slot T_N. The narrow pulse requires expanding the channel bandwidth over the bandwidth of the baseband.

At the PPM receiver, the arrival time of the pulse must be obtained, rather than the amplitude or detailed shape. The PPM detector extracts pulse time displacements, converts them to uniformly spaced voltage samples and then low-pass filters to reconstruct the baseband analog signals. Because of the narrow pulses involved, PPM requires a bandwidth expansion of the original analog signal. Pulse modulation is basically an analog modulation technique and is considered further in Chapter 2, where we also consider pulse frequency modulation (PFM), which has gained considerable popularity with fiber optic systems.

Another popular modulation technique is pulse-code-modulation (PCM). In this scheme, the analog sample is digitized so that the amplitude of the sample is represented by a digital word. The fiber optic system then uses the binary digital data to turn the light source on and off. PCM also requires an increase in channel bandwidth over that of the baseband analog signal. Digital modulation is the subject of Chapter 3.

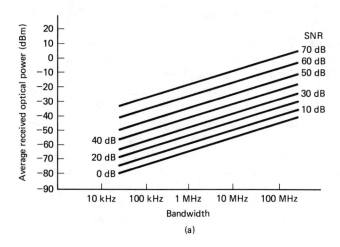

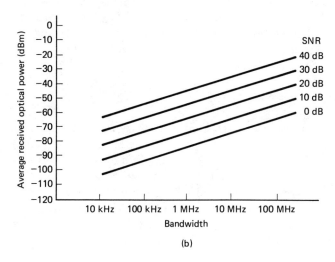

Figure 1.32 The figure of merit for analog transmission is signal-to-noise ratio (SNR). These plots of average received optical power versus bandwidth for several SNR values illustrate the capability of (a) PIN and (b) APD detectors.

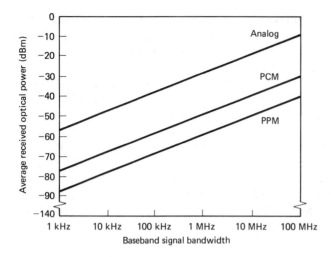

Figure 1.33 Pulse-position modulation (PPM) requires less optical power to maintain a given SNR. This advantage can significantly increase link length or enhance signal quality.

1.3.5 Signal-to-Noise Ratio

The SNR is the figure of merit used to determine how well the system transmits analog signals. The designer has considerable control over this, because the SNR is determined by the photodiode and the front end amplifier used as a receiver. This choice depends on the detector's bandwidth and the average optical power available. Performance curves for state of the art PIN and APD analog receivers are shown in Figure 1.32. The increase in sensitivity for APD receivers is readily apparent.

A comparison of analog, PPM, and PCM is of interest (Figure 1.33). To form a basis for comparison, consider a PIN detector, fix the SNR at 60 dB, and examine the required power with respect to information bandwidth. Note that to maintain a 60-dB SNR, PPM requires 30 dB less optical power than analog-intensity modulation. This has a dramatic effect on the length of the link. Similarly, 11-bit PCM provides a gain margin of approximately 18 dB. The tradeoff here is the bandwidth-expansion requirement for both PPM and PCM. Although PPM and PCM provide an increase in power margin, a limitation for both techniques occurs when we consider the allowable system dispersion (Figure 1.34). The reason for the interest in low-dispersion single-mode fibers is obvious.

1.3.6 Digital Modulation

The most straightforward scheme to use for transmitting binary digital data is simply on/off keying of the optical source. In this technique, the bit error rate (BER) replaces the SNR as the measure of performance. For example, a BER = 10^{-6} represents one error out of one million bits. The optical power required by state-of-the-art PIN and APD receivers is shown in Figure 1.35. For example, at a data rate of 10 Mbps, a PIN diode receiver requires an average received optical power of about -43 dBm. An APD receiver requires around -60 dBm for the data rate. This represents an excess margin of 17 dB.

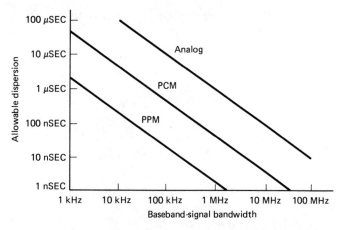

Figure 1.34 The shortcomings of PPM appear in these plots. Although PPM provides an extra power margin, it suffers with regard to allowable dispersion.

In terms of intensity modulation, we could also use a hybrid subcarrier modulation scheme. We could first digitally modulate an RF subcarrier (using frequency-amplitude or phase-shift keying, for example). The modulated subcarrier would then modulate the intensity of the optical source. Chapter 3 is devoted to the subject of digital modulation.

1.3.7 Loss Budgeting in System Design

In order to ensure satisfactory performance, the designer must budget the optical losses in the sytem. From Figures 1.33, 1.34, and 1.35 we can determine the average

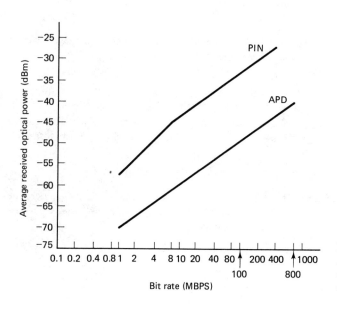

Figure 1.35 Enhanced sensitivity of the APD receiver appears when you consider the optical power required by state-of-the-art detectors.

Sec. 1.3 System Design Parameters

optical power, P_R, required by the photodetector to maintain a proper SNR or BER. The loss limit for the link is given by

$$L = P_S - P_R \qquad (1.18)$$

where

$$P_S = \text{source power}$$
$$P_R = \text{received power}$$

In a point-to-point link, we encounter input-coupling losses (ICL) between the source and the fiber, connector and/or splice losses (CSL), attenuation loss (AL) within the fiber, and output-coupling loss (OCL). Measured in decibels, the link loss budget is then

$$L = \text{ICL} + \text{CSL} + \text{AL}(D) + \text{OCL} \qquad (1.19)$$

where D is the *power-limited-distance* of the link.

Beyond this distance, D, the received power is not sufficient to maintain the required SNR or BER. If we assume that dispersion does not represent a problem, D represents the maximum repeater spacing. Generally speaking, as a safety margin, a few decibels are added to the loss budget to account for factors such as power temperature degradation. Loss budgeting for multiterminal systems is a bit more complex. We must account for such factors as coupler-tap-ratio losses, coupler insertion losses, varying distances to terminals, and dynamic range considerations. Example 1.8 illustrates link design.

Example 1.8

It is desired to transmit a 100-MHz analog signal over a distance, D, of 2 km with a received SNR of 40 dB. Use an analog intensity-modulated LED and calculate the loss budget, using typical figures.

Solution From Figure 1.32, at 1 MHz, both the PIN and the APD need about the same amount of received power at a 40-dB SNR. That is, $P_R = -35$ dBm. For cost considerations, choose the PIN diode and assume a Lambertian-emitter LED source with a total radiant power of 2 mW (+ 3 dBm). From Equation (1.18), the loss limit L is

$$L = P_S = P_R = +3 - (-35) = 38 \text{ dB}$$

Next, consider the ICL loss. If the LED's emitting area is smaller than the core, the ICL is due primarily to the NA of the fiber plus 0.2 dB for reflection loss. Choose a fiber with an NA of 0.25. Then the ICL loss is

$$\text{ICL} = 10 \log (0.25)^2 + 0.2$$
$$= 12.2 \text{ dB}$$

Now, as an estimate, let the OCL loss be 1 dB and assume three connector losses (one at each end and one at the middle). Then CSL = 3 dB. From Equation (1.19), the loss budget is

$$L = ICL + LSL + AL(D) + \text{safety}$$
$$= 12.2 + 3 + AL(D) + 1 + 3$$

Then

$$AL(D) = 18.8 \text{ dB}$$

The allowable cable loss for a 2-km link is then $18.8/2 = 9.4$ dB/km. If a higher loss fiber is considered, it will be necessary to increase the source power.

1.3.8 Signal Degradation

Noise and distortion are responsible for signal degradation. In a fiber optic system, the receiver is the major source of noise. A signal will be harmonically distorted by nonlinearities in a light source's output power-vs-drive current characteristic. In particular, an ILD has a strong nonlinearity at the lasing current threshold and requires proper biasing to avoid clipping (see Figure 1.22). It is clear that even if the photodetector receives the necessary optical power for a specified SNR or BER, nonlinear distortion can still degrade the performance. The effects of nonlinear distortion are more severe in analog systems than in digital systems. Digital systems using on/off keying can tolerate more harmonic distortion and exploit the advantages of the ILD.

As noted previously, material and modal dispersion can cause distortion. In pulse-modulated systems this is evident as intersymbol interference. For analog systems, the effect is band-limited amplitude distortion. For ILDs with spectral widths between 2 and 4 nm, this distortion component is negligible. The effect is much more pronounced with LEDs with spectral widths between 30 and 50 nm.

The shape of a signal is also distorted by the rise time of the source and detector. Rise times can range from 5 to 15 ns for LEDs and 0.1 to 2 nsec for ILDs. At the receiving end of the fiber, PINs and APDs have rise times between 1 and 4 nsec. The total system rise time includes the effects of the source, fiber, and detector. Generally speaking, a fiber's 10 to 90 percent rise time measures about 70 percent of the 3-dB modal dispersion figure (in ns/kilometer). This is true also for material dispersion. The total system rise time can be expressed as

$$T_{\text{system}} = 1.1 \sqrt{T_{\text{source}}^2 + T_{\text{modal}}^2 + T_{\text{material}}^2 + T_{\text{detector}}^2} \qquad (1.20)$$

where the Ts are the 10 to 90 percent rise times. In terms of digital systems, we can specify some guidelines for upper limits on T_{system}. For example, it should be less than 70 percent of the bit interval for nonreturn-to-zero (NRZ) formatted data and less than 35 percent for return-to-zero (RZ) data. Because T_{modal} and T_{material} depend linearly on fiber length (neglecting mode coupling), we can determine a system's dispersion-limited distance by specifying a value for T_{system} and solving Equation

(1.20) for the allowable length beyond which the system rise time specification is exceeded. System aspects leading to Equation (1.20) are discussed further in Chapter 2.

1.4 DRIVE AND RECEIVE CIRCUITS

At this point, in order that we can do some work with fiber optics, the remaining system components that need to be discussed are the transmitter and receiver circuits. Basic drive circuits for diode light sources are straightforward. Figure 1.36 illustrates two such skeleton circuits. For linear-intensity modulation [Figure 1.36(a)], current flow through the LED is controlled by the biased signal current applied to the base of the transistor. Maximum forward currents for LEDs and IDLs are usually specified. For CW operation, maximum currents are typically a few hundred milliamperes. Pulsed modulation may be accomplished by using the circuit shown in Figure 1.36(b). For on/off keying, we can use an FET to switch an LED or ILD. If an ILD with a sharp "knee" at the lasing current threshold is being used, we can use resistor R to bias the ILD just below threshold. This will increase switching speed, since the ILD will not have to begin with zero current.

1.4.1 Photodiode Receiver Circuits

Two basic photodiode receiver circuits are shown in Figure 1.37. The transimpedance amplifier illustrated in Figure 1.37(a) is designed for current sources such as PIN or APD photodiodes with the amplifier providing current-to-voltage conversion. The receiver in Figure 1.37(b) operates in a voltage gain mode. The operational amplifier amplifies the voltage developed across the load resistor, R_L. In choosing a load resistor, it is necessary to make tradeoffs between the optical sensitivity and response speed (rise time). For a given photodiode, large loads ($R_L = 5$ MΩ) reduce thermal noise current from the resistor. This lowers the NEP and allows the detection of lower light levels. Signal detection is limited by the photodiode's dark-current shot noise. On the other hand, small loads ($R_L = 50$ Ω) lower the RC time

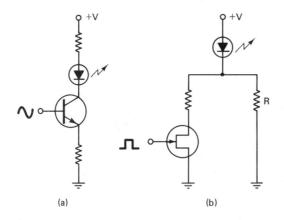

Figure 1.36 Straightforward basics suffice in the design of source drive circuits. (a) A simple transistor circuit works well for analog-intensity modulation, and you can use (b) an FET circuit for pulsed-modulation schemes.

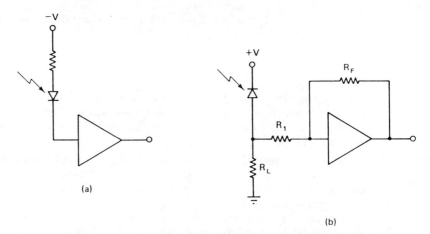

Figure 1.37 (a) Transimpedance amplifier. (b) Voltage gain receiver.

constant of the load and diode-junction capacitance. Manufacturers market hybrid modules that integrate the photodiode and preamp into one package. The responsivity and the gain–bandwidth product are specified. Temperature-compensated APD/preamp modules are also available.

In this chapter the basic principles of fiber optic systems were considered. In succeeding chapters we wish to expand on these topics sufficiently to enable the reader to acquire proficiency in designing and implementing systems. In Chapter 2 we return to the subject of analog modulation.

REFERENCES

1.1 HENRY, P. S. 1985. Introduction to lightwave transmission. *IEEE Communications* 23: 12–16.

1.2 KECK, B. D. 1985. Fundamentals of optical waveguide fibers. *IEEE Communications Magazine* 23: 17–22.

1.3 URY, ISRAEL. 1985. Optical communications. *Microwave Journal*, 24–35.

PROBLEMS

1.1. Make a plot of the number of allowed modes in a step-index fiber versus n_2 if $n_1 = 1.50$, $\lambda = 1.0$ μm, and $d = 2.5$ μm. Allow values of $1.5 < n_2 < 1.4$.

1.2. A step-index fiber has a numerical aperture of 0.16, a core refractive index of 1.45, and a core diameter of 90 μm. Find the following:
   (a) The acceptance angle, θ_c, of the fiber.
   (b) The number of modes that the fiber can carry at a wavelength of 0.9 μm.
   (c) The refractive index of the cladding.

1.3. (a) Find the numerical aperture, NA, the acceptance angle, θ_c, and the fraction of light coupled into the fiber if $n_1 = 1.5$, $n_2 = 1.4$, and the source is Lambertian. The interface to the fiber is air, $n_0 = 1.0$.

(b) Repeat (a) if the fiber is placed in a solution so that $n_0 = 1.33$.

1.4. Repeat Problem 1.3 if the angular emission of the source is described by $I = I_0 (\cos \theta)^m$. Consider values of $m = 2, 4,$ and 8.

1.5. The much narrower beam of a laser diode allows more power to be coupled into a fiber than is possible with an LED. The illuminance as a function of angle is expressed as $I(\theta) = I_0 \cos^n \theta$. A Lambertian source is characterized by $n = 1$. Using this expression, calculate the coupling efficiency into a fiber as a function of n. Assume NA = 1.

1.6. A fiber optic waveguide has a 1-mm diameter core with index 1.5 and a cladding index of 1.485. For a wavelength of 0.6 micron, find the following:
(a) maximum grazing angle for propagation
(b) number of propagating modes
(c) core diameter for a single propagating mode
(d) Repeat (a), (b), and (c) if the cladding is removed (index = 1).

1.7. Time dispersion in a fiber optic waveguide is calculated by determining the delay difference between a propagating mode with grazing angle $\theta = 0$ and the maximum grazing angle. (a) Determine an expression for delay dispersion for a guide of length L, wavelength λ, and core and cladding indices n_1 and n_2. (b) Use the parameters in Problem 1.6 to find the dispersion and bandwidth for a fiber 1 km in length.

1.8. A fiber optic cable with an attenuation of 3.0 dB/km is used in a system. The fiber is 2.8 km long and has one splice with 0.8-dB loss. The source and receiver connections each exhibit 1 dB of loss. Proper system operation requires 3 μW of received optical power at the detector. Calculate the required level of optical power from the light source.

1.9. A fiber optic waveguide has an index of refraction that is tapered from the guide center according to the following relation:

$$n(r) = n_0 \left[1 - \Delta \left(\frac{r}{d/2} \right)^2 \right]$$

where r is the radial distance, d is the core diameter, and Δ is the taper rate. For this taper rate, the maximum grazing angle θ is given by

$$\cos \theta = \left[\frac{1 - 3\Delta}{1 - \Delta} \right]^{1/2}$$

Derive the time delay dispersion.

1.10. A GaAlAs LED radiates at 0.85 μm with a power of 0.5 mW when a current of 150 ma flows through the diode and a bias voltage of 1.5 V is applied. Find the overall efficiency.

1.11. Twenty ma flows through a silicon and a zinc-doped GaAs diode when 1.2 volts is applied at 25°C, and 17 ma flows at 100°C when 1.1 V is applied. Find the power emitted at both temperatures from each diode if the silicon-doped LED is 10 percent efficient at 25°C and 5 percent efficient at 100°C. The zinc-doped LED is one-fourth as efficient.

2

Analog Modulation

2.0 INTRODUCTION

In today's communication technology, fiber optics is generally regarded as a medium for transmitting digital signals. It is also capable of handling analog signals. The most powerful aspect of optical communication, and the one so far least exploited for either analog or digital, is the tremendous bandwidth available at optical frequencies. A wavelength of 1 µm corresponds to 3×10^{14} Hz and as noted in Chapter 1, a single 16-GHz channel corresponds to a mere 3.3×10^{-6} µm of wavelength spread. Optical communication systems with bandwidths exceeding 1 GHz are now available.

2.1 OPTICAL CHANNEL

From Chapter 1 we know that in order to transmit information optically, one needs an optical channel, a transmitter (modulator), and a detector (receiver). The simplest method of transmitting optical energy is to allow it to propagate through free space. Optical beams can be made to propagate with very low divergence because of the very short wavelengths involved. For example, at 1-µm wavelength, a 1-meter antenna (telescope) would have a beam divergence on the order of 1 microradian. In the atmosphere, optical absorption and scattering severely limit the use of freely propagating beams. This is not the case in the vacuum of free space. In this setting, optical beams can provide significant performance advantages over microwaves such as

- Small antenna (telescope) size
- Low sidelobes
- Low power requirements

Two uses for optical channels are illustrated in Figure 2.1. In this illustration, two satellites in geosynchronous orbit are in communication, and data from a low-altitude satellite are being relayed to a satellite in geosynchronous orbit. A second optical channel is the optical fiber discussed in Chapter 1. The great advantage of optical fiber is that it is low-loss, very wideband, lightweight, and quite flexible. In order to understand how an optical channel performs, it is necessary to understand the physical limits to optical detection. We wish to relate these limits to the detection of optical radiation.

2.1.1 Detection of Optical Radiation

At microwave frequencies, the quantum nature of the electromagnetic field is not readily apparent. For example, at a wavelength of 10 cm (30 GHz), the energy of a single photon is only 0.1 meV. This is small, indeed, compared to the thermal energy kT of 26 meV at room temperature. Under these conditions, it takes many photons to rise above the surface of the sea of thermal photons. Thus, the granular nature of the received electromagnetic energy is not noticeable. At the same time, at a wavelength of 10 cm, a single watt of power carries with it roughly 10^{24} photons

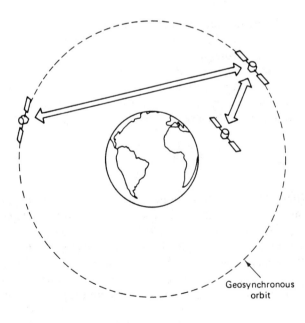

Figure 2.1 The large bandwidth of the optical channel can speed communications between satellites in space. This diagram shows communication between a low-flying satellite and one in geosynchronous orbit or between two satellites, both of which are in geosynchronous orbit.

per second. We see, then, that a single photon contains very little energy at this frequency.

Now, at an optical wavelength of 1 μm, the photon has an energy of 1.2 eV. This is far above the thermal background, and photons are far less plentiful for a given power level. The randomness of the arrival of these energetic photons at the photodetector generates shot noise in the detected signal. In order to calculate the effect of shot noise, we first represent the electric field of the incident optical signal by

$$e_s(t) = \mathrm{Re}\, [E_s e^{i\omega_s t}] \tag{2.1}$$

From Chapter 1, we know that the photodetector operates on the principle that the arriving photons generate free carriers. This results in a current flowing through the external circuit (see Figure 1.24). The signal current, $i_s(t)$, is proportional to the incident optical power, P_s, and the power is proportional to E_s^2. The signal current generates shot noise current, i_{sh}, given by

$$i_{sh}^2 = \langle i_{sh}^2 \rangle = 2e i_s B \tag{2.2}$$

where

$B =$ the bandwidth

$e = 1.6 \times 10^{-19}$ coulombs

$i_s =$ signal current

$\langle i_{sh}^2 \rangle =$ mean square value

For a detector with quantum efficiency, η, we have

$$i_s = \frac{\eta P_s e}{hf} \tag{2.3}$$

where

$h =$ Planck's constant, 6.625×10^{-34} J·sec

$f =$ optical frequency

The quantity hf represents the energy in a photon in electron volts. The ratio of the electrical signal power to the electrical noise power is the SNR:

$$\frac{i_s^2}{i_{sh}^2} = \frac{\eta P_s}{2hfB} \tag{2.4}$$

The SNR is proportional to the received optical power. Equation (2.4) represents the SNR at the output of the photo-detector and is theoretically the best signal-to-noise ratio that can be achieved. In practice, however, this limit is usually not achieved. First of all, lasers generally exhibit intensity fluctuations greater than those associated with shot noise. These fluctuations are a fixed percentage of the received laser power. Therefore, this effect is referred to as *relative intensity noise*. This places an upper limit on the SNR of the channel. Now, if the signal is attenuated as it propagates through a long fiber, the SNR, due to shot noise, will be worse than the value predicted from relative intensity noise. In practice, the first noise source to exceed the relative intensity noise is the thermal noise of the receiver. In order to achieve theoretical shot noise limited noise performance from an attenuated signal, heterodyne detection can be employed.

Heterodyne detection, a coherent optical transmission technique, is discussed in Chapter 7. It is of interest to introduce the technique at this point in connection with shot noise. Consider once again that we have not only an incident electric field but also a local electric field given by

$$e_{LO}(t) = \text{Re}\,(E_{LO}e^{j\Omega_{LO}t}) \tag{2.5}$$

where a separate local oscillator laser is available. Both electric fields are incident on the photodetector. The photodetector current, $i(t)$, will now be proportional to

$$|E_s e^{\Omega_s t} + E_{LO} e^{\Omega_{LO} t}|^2 = E_s^2 + E_{LO}^2 + 2 E_s E_{LO} \cos(\Omega_s - \Omega_{LO})t$$

The local oscillator power is chosen to be much larger than the power in the signal we are trying to detect. Thus, the shot noise is generated by the average detected current, and this average current is proportional to the local oscillator power, P_{LO}. We identify the signal current, $i_s(t)$, with the difference frequency term, $\Omega_s - \Omega_{LO}$. Note that it is proportional to the product of the electric fields, $[P_s P_{LO}]^{1/2}$. The SNR is still proportional to P_s, as it is for ordinary shot noise, but the signal has been increased by a factor of $[P_{LO} P_s]^{1/2}$, thus allowing it to rise above the thermal noise of the amplifier.

Heterodyne detection techniques are presently still in the laboratory stage primarily because of difficulties inherent in stabilizing the source and local oscillator frequencies. When this problem is resolved, it will be possible to achieve essentially quantum-limited performance in an optical channel.

2.1.2 Direct Detection Optical Communication Receiver

A receiving system that views an atmosphere's background also collects undesirable background radiation that falls within the spatial and frequency ranges of the detector. This radiation is processed with the signal and represents an overall degradation in performance. We can model the direct-detection receiver as shown in Figure 2.2. The incident field results in a current through the P–N junction of the photodiode. In addition, the detected signal produces a shot noise process at the detector output [see Equation (2.2)]. The shot noise process is governed by the received field intensity arising from both the signal and the background radiation collected in the optical front end. This combination produces beat frequencies. Obviously, in a fiber optic link, the background radiation will be excluded. Prior to receiver postdetection processing, circuit (thermal) noise is added to the detector output.

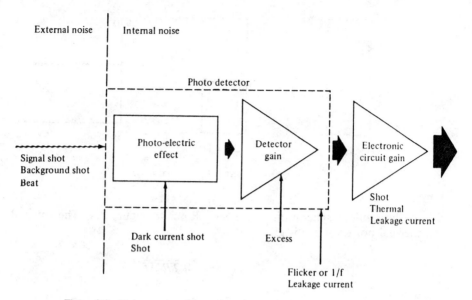

Figure 2.2 Noise sources (direct detection optical communication receiver).

Two optical receiver designs were illustrated in Figure 1.36. As a basis for noise discussion, one of these receivers is illustrated in Figure 2.3. Typically for wide-band applications only shot noise and thermal noise need be considered in calculating receiver performance. Remember, for an attenuated signal, these noise sources overtake relative intensity noise. The total noise current, i_{nt}, into the preamplifier is the rms value of the noise current from the photodiode and the thermal noise current from load resistor, R_L.

2.1.2.1 Noise Considerations. Thermal noise is caused by the random thermal motion of charged carriers. This is in contrast to shot noise, which is caused

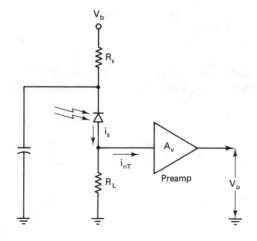

Figure 2.3 Optical receiver.

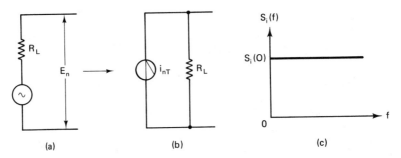

Figure 2.4 Thermal noise.

by the random generation of carriers. Refer to Figure 2.4. The rms value of the thermal noise voltage is given by

$$E_n = \sqrt{4kTR\,\Delta f} \qquad (2.6)$$

where

k = Boltzmann's constant: 1.38×10^{-23} W-S/deg

T = temperature, K

Δf = bandwidth, Hz

If the Thevenin equivalent circuit is converted to a Norton's equivalent circuit [Figure 2.4(b)], the thermal noise current for resistor R can be expressed as

$$i_{th} = \frac{\sqrt{4kT\,\Delta f}}{R}$$

or

$$i_{th}^2 = \langle i_{th}^2 \rangle = \frac{4kT\,\Delta f}{R} = \frac{4eV_T\,\Delta f}{R} \qquad (2.7)$$

where $e = 1.6 \times 10^{-19}$ amps/s is the charge of an electron and $V_T = kT/e$. The thermal noise from the resistor has a uniform spectral distribution, $S_i(0)$, which may be expressed as

$$S_i(0) = \frac{4eV_T}{R} \qquad (2.8)$$

This is indicated in Figure 2.4(c). At room temperature, $T \approx 300°\text{K}$ and $V_T = 0.025$ V.

Example 2.1

Calculate the thermal noise current at room temperature for:

(a) $R = 1\ M\Omega$, $\Delta f = 100\ kHz$

(b) $R = 1\ M\Omega$, $\Delta f = 10\ MHz$

(c) $R = 10\ \Omega$, $\Delta f = 10\ MHz$

Solution

(a) $i_{th} = \dfrac{\sqrt{(4)(1.38 \times 10^{-23})(300)(10^7)}}{10^6} = 0.04\ na$

(b) $i_{th} = \dfrac{\sqrt{(4)(1.38 \times 10^{-23})(300)(10^7)}}{10^6} = 0.4\ na$

(c) $i_{th} = \dfrac{\sqrt{(4)(1.38 \times 10^{-23})(300)(10^7)}}{10} = 126\ na$

Note from this example that the thermal noise current increases with increasing bandwidth Δf and decreasing value of the resistance R.

Refer to Figure 2.5. Shot noise produced by a photodiode carrying a current i can be presented by a current source with a mean square value given by Equation (2.2). The shot noise has a uniform spectral distribution given by

$$S_i(0) = 2eI \tag{2.9}$$

where I is the average or dc value of the diode current i.

Example 2.2

Calculate the shot noise current for:

(a) $I = 10\ \mu a$, $\Delta f = 10\ MHz$

(b) $I = 1\ \mu a$, $\Delta f = 10\ MHz$

Solution

(a) $i_{sh} = \sqrt{(2)(1.6 \times 10^{-19})(10 \times 10^{-9})(10^7)} = 0.18\ na$

(b) $i_{sh} = \sqrt{(2)(1.6 \times 10^{-19})(10^{-6})(10^7)} = 1.8\ na$

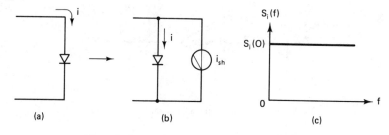

Figure 2.5 Shot noise produced by a photodiode.

The mean square value i_{nT}^2 of the total noise is the sum of the mean squares of the individual noise components. Thus,

$$i_{nT}^2 = \langle i_{nT}^2 \rangle = i_{sh}^2 + i_{th}^2 = \left(2eI + \frac{4eV_T}{R}\right)\Delta f \qquad (2.10)$$

$$= 2e\left(I + \frac{2V_T}{R}\right)\Delta f$$

For $\dfrac{2V_T}{R} = 50 \text{ mV}/R \gg I$, Equation (2.10) reduces to

$$i_{nT}^2 \approx \frac{4eV_T}{R}\Delta f \qquad (2.11)$$

In other words, the shot noise contribution can be neglected, and only thermal noise need be taken into consideration.

The optical receiver can be modeled as shown in Figure 2.6. Equation (2.4) expressed the SNR at the output of the photodiode. With the thermal noise contribution, the SNR at the input to the preamplifier is equal to

$$\text{SNR} = \frac{i_s^2}{i_{nT}^2} \qquad (2.12)$$

where

$$i_{nt} = \sqrt{i_{sh}^2 + i_{nt}^2}$$

and

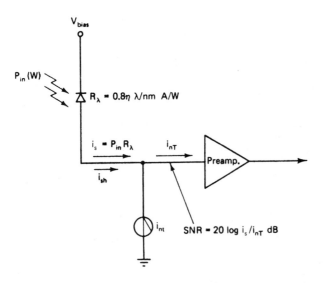

Figure 2.6 Optical receiver noise model.

50   Analog Modulation   Chap. 2

$$i_s = P_s R_\lambda$$

$$= P_s \left(0.8 \text{ n} \frac{\lambda}{\text{nm}}\right) \text{ amperes, photocurrent}$$

We noted previously that the minimum detectable signal depends on the noise level. The shot noise depends on the signal level, whereas the thermal noise contribution is independent of I. In order to make i_{th} as low as possible, it is necessary to make the input resistance of the amplifier as high as possible. However, high data rates require amplifiers with wide bandwidth and fast response. This requirement places an upper limit on the time constant, $\tau = RC$, of the amplifier. Also, the input capacitance C must be kept low: <2 pF. This places an upper limit on the value of R. The equivalent circuit for Figure 2.3 is shown in Figure 2.7, and some typical circuits for the receiver input are shown in Figure 2.8.

2.2 BASEBAND ANALOG MODULATION AND DETECTION

For baseband intensity modulation, the transmitted optical signal is described by

$$P(t) = P_0[1 + mf(t)] \tag{2.13}$$

where

P_0 = average optical power

$m \leq 1$ = modulation index

$f(t)$ = normalized signal with $|f(t)|$ max = 1, $\langle f(t) \rangle = 0$, and $\langle f^2(t) \rangle \approx \frac{1}{2}$ (the symbol $\langle \ \rangle$ means "average")

The instantaneous photocurrent i_p is proportional to the received optical signal. Thus,

$$i_p = I_0[1 + mf(t)] \tag{2.14}$$

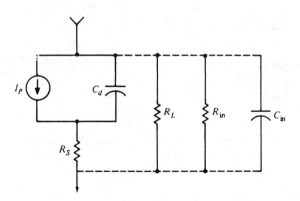

Figure 2.7 Input preamplifier equivalent circuit.

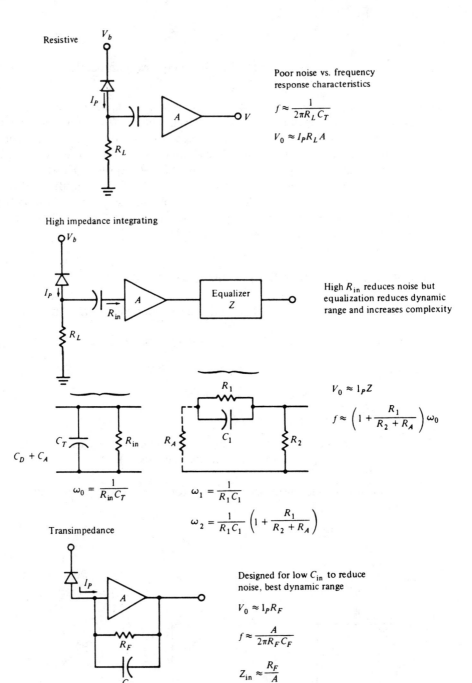

Figure 2.8 Receiver electronics: resistive, high-impedance integrating, and transimpedance.

and $\langle i_p \rangle = I_0$, the average photocurrent. The mean square value of the photocurrent associated with the signal is

$$\langle i_s^2 \rangle = m^2 I_0^2 \langle f^2(t) \rangle \qquad (2.15)$$
$$\approx \tfrac{1}{2} m^2 I_0^2$$

Refer to Figure 2.6. The signal-to-noise ratio, SNR, is equal to the ratio of the signal power to the noise power, or equivalently, to the ratio of the mean square value of the signal current to the mean square value of the noise current. That is,

$$\text{SNR} = \frac{\langle i_s^2 \rangle}{\langle i_{nT}^2 \rangle} = \frac{1}{2} \frac{m^2 I_0^2}{2e\left(I_0 + \dfrac{2V_T}{R}\right)f} \qquad (2.16)$$

From Equation (2.16), it is clear that the maximum SNR occurs when the modulation index $m = 1$. For a preamplifier with –6-dB/octave roll-off, the noise bandwidth B_n is given by

$$B_n = \frac{1}{4RC} \qquad (2.17)$$

This is characteristic of a transimpedance amplifier, as shown in Figure 2.8. If the baseband-width of the signal $f(t)$ is equal to the –3-dB bandwidth of the amplifier, Equation (2.17) may be written as

$$B_n = \frac{W_{-3\,\text{dB}}}{4} = \frac{\pi}{2} f_{-3\,\text{dB}} \qquad (2.18)$$

and the maximum SNR can be expressed as

$$\text{SNR}_{\max} = \frac{I_0^2}{2\pi f_{-3\,\text{dB}}\, e(I_0 + 2V_T/R)} \qquad (2.19)$$

where R is the equivalent feedback resistance of a transimpedance preamplifier or the input resistance of an amplifier without feedback.

To maximize the SNR we need to make R as large as possible (to reduce thermal noise). However, the required –3-dB bandwidth of the amplifier imposes an upper limit on the value of R. Thus, if C is the effective feedback capacitance across R, we require that $R = \dfrac{1}{2\pi f_{-3\,\text{dB}} C}$. We can rewrite Equation (2.19) as

$$\text{SNR}_{\max} = \frac{I_0^2}{2\pi f_{-3\,\text{dB}}\, e(I_0 + 4\pi V_T f_{-3\,\text{dB}} C)} \qquad (2.20)$$

From this equation, for any desired SNR, we can determine the minimum detectable signal current I_0 and the power level P_0 at the receiving end; specifically,

$$I_0 = \pi f_{-3\,\text{dB}}\, e\, \text{SNR}_{\max} \left(1 + \sqrt{\frac{8 V_T C}{e\, \text{SNR}_{\max}}}\right) \qquad (2.21)$$

The optical power is related to the photocurrent by

$$\eta P_0 = \frac{1.24}{\lambda/\mu m} I_0 \text{ W/A} \qquad (2.22)$$

$$= \frac{1.24\ \mu m}{\lambda} I_0 \text{ W/A}$$

The lower the value of C, the lower the required signal level. For well-designed transimpedance amplifiers, C is kept below 0.5 pF.

Example 2.3

It is desired to transmit black-and-white television (bandwidth = 4.5 MHz) using red light ($\lambda = 6 \times 10^{-7}$ m). Assuming a photodiode quantum efficiency of 100 percent, calculate the minimum detectable signal power. Assume an effective feedback capacitance across R of 0.5 pF and a required SNR of 47 dB (5×10^4).

Solution

$$I_0 = 1.601 \times 10^{-19})(3.14)(4.5 \times 10^6)(5 \times 10^4)$$

$$\times \left[1 + \sqrt{1 + \frac{(8)(0.025)(0.5 \times 10^{-12})}{(1.601 \times 10^{-19})(5 \times 10^4)}}\right]$$

$$= 528.83 \text{ nA}$$

$$P_0 = \frac{1.24 \times 10^{-6}}{6 \times 10^{-7}} (528.83 \times 10^{-9})$$

$$= 1.09\ \mu m$$

2.2.1 APD Detector

We discussed the characteristics of avalanche photodiodes in Chapter 1. An APD diode provides internal gain and generates on the average G_0 photoelectrons for each primary photoelectron. Thus, the photocurrent is enhanced by the gain factor G_0. That is,

$$i_p \text{ (APD)} = G_0 i_p \text{(PIN)} \qquad (2.23)$$

The avalanche gain G_0 can be as high as 200. Thus, we see that it is reasonable to expect that an APD detector will yield a considerable reduction in input power level requirements. This, however, is not always the case. Because of the randomness of the avalanche multiplication process, the shot noise of the APD increases with increasing G_0 by a larger factor than the mean square value of the signal current. The following conditions apply:

$$\langle i_p^2(\text{APD})\rangle = G_0^2 \langle i_p^2(\text{PIN})\rangle \tag{2.24}$$

and

$$\langle i_{\text{sh}}^2(\text{APD})\rangle = \langle G^2\rangle \langle i_{\text{sh}}^2(\text{PIN})\rangle \tag{2.25}$$

where $\langle G^2\rangle = G_0^2 F(G_0)$. Now, $F(G_0) > 1$ and is the excess noise factor. For silicon avalanche photodiodes, it is given approximately by

$$F(G_0) \approx \sqrt{G_0} \tag{2.26}$$

Thus,

$$\langle i_{\text{sh}}^2\rangle = G_0^2 \sqrt{G_0}(2eI_e \,\Delta f) \tag{2.27}$$

From this expression we see that an APD would be of little use if only shot noise were present. It is the presence of thermal noise that often dictates its use as a photodetector.

The mean square value of signal current and noise current is given by

$$\langle i_s^2\rangle = m^2 G_0^2 I_0 \langle f^2(t)\rangle$$
$$\approx \tfrac{1}{2} m^2 G_0^2 I_0^2 \tag{2.28}$$

$$\langle i_{\text{nt}}^2\rangle = \langle i_{\text{sh}}^2\rangle + \langle i_{\text{th}}^2\rangle$$
$$= 2e\left(G_0^2 \sqrt{G_0} I_0 + \frac{2V_T}{R}\right)\Delta f \tag{2.29}$$

For an equivalent noise bandwidth, $B_n = \pi/2 f_{-3\,\text{dB}}$, we have

$$\text{SNR} = \frac{\langle i_s^2\rangle}{\langle i_{nT}^2\rangle}$$

$$= \frac{m^2 I_0^2}{2\pi e\left(\sqrt{G_0}\, I_0 + \dfrac{2V_T/R}{G_0^2}\right) f_{-3\,\text{dB}}} \tag{2.30}$$

Examining Equation (2.30), we see that the effect of the avalanche gain is to increase the shot noise by the factor G_0 and to decrease the thermal noise by the factor G_0^2. Thus, we can conclude that the APD is useful in applications where the shot noise is small compared to the thermal noise. Specifically, this involves modulation schemes that result in low signal-level requirements. If we set $m \approx 1$, the signal-to-noise ratio is given by

$$\text{SNR} = \frac{I_0^2}{2\pi e \sqrt{G_0} I_0 + \dfrac{2V_T/R}{G_0^2} f_{-3\,\text{dB}}} \tag{2.31}$$

For $R = \dfrac{1}{2\pi f_{-3\,\text{dB}} C}$, this expression becomes

$$\text{SNR} = \frac{I_0^2}{2\pi e \left(\sqrt{G_0} I_0 + \dfrac{4\pi V_T f_{-3\,\text{dB}}}{G_0^2} \right) f_{-3\,\text{dB}}} \tag{2.32}$$

Referring to Equation (2.32), we see that the SNR varies with the factor G_0. Maximum SNR occurs when the denominator is minimum. It can be shown that the maximum SNR is given by

$$\text{SNR}_{\text{max}} = \frac{2 I_0}{5\pi e f_{-3\,\text{dB}}} \left(\frac{I_0}{16\pi V_T C f_{-3\,\text{dB}}} \right)^{1/5} \tag{2.33}$$

We can solve this equation for the minimum required signal current I_0:

$$I_0 \geq \frac{5}{2} \pi \left(\frac{V_T C}{5e} \right)^{1/6} \text{SNR}_{\text{max}}^{5/6} e f_{-3\,\text{dB}} \tag{2.34}$$

If we assume a quantum efficiency $\eta = 1$ and $\lambda = 0.8\ \mu\text{m}$, the minimum required optical power P_0 of the receiver is

$$P_0 = \frac{1.24\ \mu\text{m}}{0.8\ \mu\text{m}} I_0 = 1.55 I_0 \tag{2.35}$$

A comparison of these results with a PIN diode is given in Table 2.1 with $C = 0.5$ pF.

TABLE 2.1

SNR_{max}	$P_0(\text{APD})/P_0(\text{PIN})$
10	−13.2 dB
10^2	−9.9 dB
10^3	−6.6 dB
10^4	−3.6 dB
10^5	−1.1 dB

Note that only when the SNR is below 30 dB does the APD afford a meaningful advantage for analog amplitude-modulated signals.

2.3 MULTIPLE-CHANNEL MODULATION

When multiple information channels are sent over a communication link, a distinct carrier frequency is assigned to each channel. Normal amplitude modulation, with carrier and both sidebands, requires a high power level and wide bandwidth and is therefore unsuitable for a fiber link. Amplitude modulation with suppressed carrier is equally unsuitable. We therefore consider suppressed-carrier single-sideband intensity modulation (SCSSB-IM).

For N analog SCSSB channels, the transmitted optical signal is given by

$$P(t) = P_o[1 + M_1 f_1(t_i W_{c_1}) + M_2 f_2(t_i W_{c_2}) + \cdots + M_N f_N(t_i W_{c_N})] \tag{2.36}$$

where

P_o = average optical power

W_{cj} = carrier frequency of the jth channel

$f_j(t; W_{cj})$ = the single sideband signal of the jth channel, with

$\langle f_j(t; W_{cj}) \rangle = 0$; $|f_j(t; W_{cj})|_{\max} = 1$, and $\langle f_j^2(t; W_{cj}) \rangle \approx \frac{1}{2}$

M_j = modulation index of the jth channel, with

$$\sum_{j=1}^{N} M_j \leq 1$$

The photocurrent is proportional to the received optical power. Thus,

$$i_p(t) = I_o[1 + M_1 f_1(t; W_{c_1}) + M_2 f_2(t; W_{c_2}) + \cdots + M_N f_N(t; W_{cN})] \tag{2.37}$$

where the average photocurrent is

$$\langle i_p(t) \rangle = I_0 \tag{2.38}$$

and the mean square value of the photocurrent associated with the jth signal is given by

$$\langle i_{s,j}^2 \rangle = M_j^2 I_o^2 \langle f_j^2(t; W_{cj}) \rangle \approx \tfrac{1}{2} M_j I_o^2 \tag{2.39}$$

We can express the signal-to-noise ratio of the jth channel as

$$\mathrm{SNR}_j = \frac{M_j^2 I_o^2}{2\pi e(I_o + 2V_T/R) f_{-3\,\mathrm{dB}}} \tag{2.40}$$

where $f_{-3\,\mathrm{dB}}$ is the basebandwidth of the jth signal and R is the equivalent feedback resistance of a transimpedance amplifier.

If we assume that all the channels have the same bandwidth and the same modulation index, we can express the SNR for N channels as

$$\text{SNR} = \frac{I_0^2/N^2}{2\pi e\left(I_0 + \dfrac{2V_T}{R}\right)f_{-3\,\text{dB}}} \tag{2.41}$$

where

$$\sum_{1}^{N} M_j = NM_j \leq 1$$

has been used, so $M = 1/N$.

For N channels, each of bandwidths $f_{-3\,\text{dB}}$, separated by a spectral width $kf_{-3\,\text{dB}}$, the preamplifier must have a bandwidth given by

$$B_{-3\,\text{dB}} \geq (k + 1)Nf_{-3\,\text{dB}} \tag{2.42}$$

and RC product given by

$$RC = \frac{1}{2\pi B_{-3\,\text{dB}}} \leq \frac{1}{2(k+1)Nf_{-3\,\text{dB}}} \tag{2.43}$$

The signal spectrum is shown in Figure 2.9. To maximize the SNR, we make R as large as possible (minimize thermal noise). We can then express the maximum SNR as

$$\text{SNR}_{\text{max}} = \frac{I_0^2/N^2}{2\pi e[I_0 + 4\pi(k+1)V_T CNf_{-3\,\text{dB}}]f_{-3\,\text{dB}}} \tag{2.44}$$

and the minimum detectable value of I_0 is given by

$$I_0 = \left[\sqrt{1 + \left(\frac{e}{16V_T C}\right)\frac{N}{k+1}\text{SNR}_{\text{max}}} + \sqrt{\left(\frac{e}{16V_T C}\right)\frac{N}{k+1}\text{SNR}_{\text{max}}}\right] \tag{2.45}$$

$$\times \sqrt{4\pi^2\left(\frac{V_T C}{e}\right)(k+1)N\text{SNR}_{\text{max}}\,eNf_{-3\,\text{dB}}}$$

Examining Equation (2.45), we see that the minimum detectable current increases at a rate proportional to $N^{3/2}$. In addition, a slight nonlinearity of the light source, LED or ILD, will result in cross modulation and cross talk interference. Analog amplitude modulation is therefore unsuitable for voice communications when the number of channels is large.

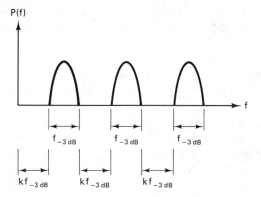

Figure 2.9 Signal spectrum.

2.3.1 Analog Modulation Methods

We see from the preceding section that to transmit analog signals optically requires more than just the proper selection of light sources and optical fibers. How the signal is modulated is also extremely important, especially for transmitting large amounts of power over long distances. For long-range transmission, we must be able to drive the diode current from zero up to the highest value. In analog systems, some type of angle modulation of diode current—to modulate the signal on the time axis—will deliver the necessary long-range power. Amplitude modulation works only where moderate transmitting power and small end-to-end losses in the fiber can be tolerated. As indicated in Figure 1.30, with a sine-wave or pulse carrier, we can use any of the usual time-axis modulation methods such as phase modulation (PM), frequency modulation (FM), pulse position modulation (PPM), or pulse frequency modulation (PFM).

If the transmission bandwidth is about twice as wide as the signal bandwidth, PM or FM yields the longest range. Moreover, FM provides a higher SNR than PM for video signals, because the amplitude density of random noise after demodulation rises linearly with frequency. In addition, the gain due to weighting of the noise power density adds another 3 dB over the gain available with PM. When the ratio of transmission bandwidth to signal bandwidth gets much greater than two, PM techniques are preferred over FM, especially if a laser diode is being used to transmit video. Pulse modulation can also be used with LEDs when transmission bandwidth is much greater than the signal bandwidth.

We can express the carrier-to-noise ratio (CNR) in an FM channel as

$$\text{CNR} = \frac{\langle i_p^2 \rangle R}{FkTB} \tag{2.46}$$

where

$\langle i_p^2 \rangle$ = mean square value of the photocurrent received in a given channel

R = load impedance

F = preamplifier noise figure

In this expansion, kTB is the noise due to the source; thus $FkTB$ is the equivalent input noise. This equation is valid if the laser intensity noise and the noise generated by intermodulation products are negligible. As we noted previously, under most conditions the receiver noise will be dominated by the thermal noise of the amplifier, and the other noise sources are negligible. It is common practice in the literature to use the term CNR instead of SNR to signify that the recovered signal is at an RF level and not baseband.

An analog transmission system can use the FM format developed for satellite video transmission. The channel-to-channel separation is 40 MHz, the receiver bandwidth is 30 MHz, and the peak frequency deviation is 8.8 MHz. By taking advantage of preemphasis/deemphasis noise reduction techniques (as in Dolby in tape recorders) and using the National Television Standards Committee (NTSC) weighting factor, it has been found that a studio-quality video signal with a 45-dB weighted signal-to-noise ratio requires a CNR of 16.5 dB.

Example 2.4

Calculate the required rms photocurrent per channel for FM in order to have a 16.5-dB CNR. Assume a preamplifier noise figure of 3 dB, a 50-ohm load impedance and room temperature. Channel bandwidth is 30 MHz.

Solution

$$\langle i_p^2 \rangle = \frac{(FkTB)\text{CNR}}{R}$$

$$= \frac{(2)(1.38 \times 10^{-23})(300)(3 \times 10^7)(44.66)}{50}$$

$$= 22.10^7 \times 10^{-14} \text{ (amp)}^2$$

$$i_p = 0.47 \ \mu\text{A}$$

The CNR of 16.5 dB required for studio-quality FM video transmission is the same as that required for digital transmission. Thus, contrary to widespread belief, analog video signals are no more susceptible to noise, nonlinearities, or intermodulation effects than digital signals. This is true provided that a wide-deviation FM format is used rather than the more common amplitude modulation used for cable and over-air video broadcasting. FM transmission requires a receiver bandwidth about one-fourth as wide as that required for 100 Mbit/s digital transmission. Consequently, the required rms value of photocurrent for FM is about half the magnitude. The modulation characteristics of a 60-channel FM subcarrier system is shown in Figure 2.10 for an 18-kilometer link.

In this particular system, for an 18-kilometer transmission distance, a modulation depth of only 2 percent per channel results in a 56-dB weighted signal-to-noise ratio. Since the required carrier-to-noise ratio is only 16.5 dB, the FM analog signal is insensitive to intermodulation products and laser noise.

As a matter of note, however, work is also continuing on development of AM systems. Since the receiver requires an AM signal, an FM-to-AM conversion is necessary for video systems. Efforts are being made to remove the nonlinearities

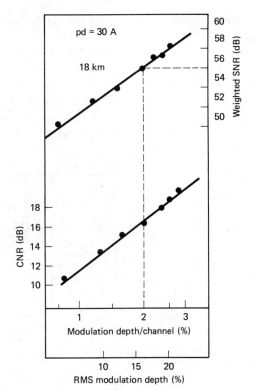

Figure 2.10 Sixty-channel FM subcarrier system.

that contribute to cross talk in AM systems. The best AM systems reported to date are still three dB less in performance than FM.

2.4 DESIGN OF FREQUENCY DIVISION MULTIPLEXED FIBER OPTIC LINKS

A block diagram of an FDM (frequency division multiplexed) analog transmission fiber optic link is illustrated in Figure 2.11. We have previously considered elements of this system. The optical transmitter accepts the individual RF FDM analog inputs

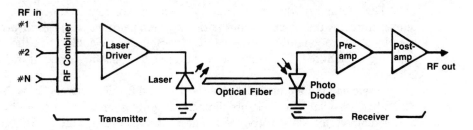

Figure 2.11 FDM analog transmission fiber optic link. (From *RF Design*, April 1987, with permission.)

and provides the necessary signal conditioning to drive the laser diode. Major optical transmitting functional components are illustrated in Figure 2.12. The RF combiner sums the multiple analog inputs that are to be transmitted. The RF levels of each carrier should be equalized prior to the optical transmitter. This will ensure optimum noise and distortion performance for each carrier. A slope compensation stage may be provided after the combiner stage to adjust for normal cable slope (frequency roll-off) effects.

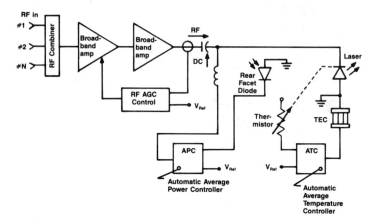

Figure 2.12 RF analog optical transmitter. (From *RF Design,* April 1987, with permission.)

AGC-controlled broadband amplifiers provide the necessary signal level to drive the laser diode. In order to optimize system noise and distortion performance, the RF drive level to the laser diode must be precisely controlled. DC bias is applied to the laser to provide a linear operating point (see Figure 1.22). This DC bias current determines the average optical output power of the laser diode. This is typically 0.5 mW for single-mode 1300 nm lasers. The power of laser diodes is sensitive to both aging and changes in temperature. In order to preserve a constant average optical output power, two control circuits are commonly provided in a laser transmitter: a laser temperature controller and an automatic optical power controller. A photodiode monitors the rear facet of the laser to sample the transmitted optical power. This information is then used to control the laser DC bias current or, in other words, to maintain a constant average optical power.

The lifetime of the laser diode is adversely affected by operating at higher temperatures. A thermistor monitors the laser temperature, and this information is used by a control circuit to drive a TEC (thermal electric cooler) to which the laser heatsink is mounted. This maintains the laser at constant temperature, typically 20°C.

Single-mode 1300-nm fiber cable is preferred for multichannel analog systems. This fiber is available with a core diameter of only 9 μm with an overall cladding/buffer diameter of 950 μm (typical). These fibers, in turn, may be assembled into cable assemblies, which provide multiple fibers and also, strain relief and jacket options. The fiber cable is available in lengths up to several kilometers per reel. For distances longer than several kilometers, the fibers are generally fusion-spliced to minimize path insertion losses. Typically, splices for single-mode fibers have losses of only a few tenths of a dB. Connectors for single-mode fibers have losses on the order of 0.5 dB. Generally, connectors are provided on the transmitter and receiver for convenience in servicing the equipment.

From previous discussions, we know that the primary function of the analog optical receiver is to reconvert the light power into an RF signal. Furthermore, this must be accomplished with a minimum contribution of noise and distortion. A block diagram of the optical receiver is shown in Figure 2.13. The optical detector commonly employed for 1300-nm analog applications is either an LnGaAs PIN diode or a Ge avalanche photodiode. The major distinction is that the avalanche diode has a gain of approximately 10, whereas the PIN diode does not exhibit gain. We discussed criteria for selection of diode (PIN or avalanche) in Section 2.2.1. Current from the photodiode drives a transimpedance preamp, which provides high input sensitivity and transforms the input current into an output voltage (see Figure 2.8). These preamps are available as DIP packaged devices with fiber pigtails attached. The preamp is followed with an AGC-controlled postamplifier, which provides sufficient gain to obtain unity system gain for the entire fiber optic link. The AGC maintains a constant output level independent of optical input power. This compensates for changes due to such things as fiber resplicing and fiber loss versus temperature.

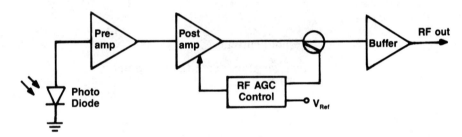

Figure 2.13 RF analog optical receiver. (From *RF Design,* April 1987, with permission.)

2.4.1 Key System Performance Parameters

In the preceding section, a general discussion of an analog fiber optic system was provided. We wish now to become specific. The key performance parameters used to characterize fiber optic analog systems are as follows:

- Carrier-to-noise ratio (CNR)
- Bandwidth
- Distortion

The CNR obtained at the receiving end of the fiber optic system ultimately determines the baseband (video) signal-to-noise performance. The output of an optical receiver is an RF carrier. Thus, CNR is preferred over system SNR as a fiber optic link parameter. Also, since analog video may be transmitted via VSB-AM (vestigial sideband), FM, or other modulation technique, for a given CNR, the resultant baseband video will be a function of the modulation approach. It is well known that wideband FM provides higher signal-to-noise ratios at the sacrifice of greater channel bandwidths.

The optical transmitter impacts the CNR in two ways:

- Setting of a particular modulation depth
- Inherent laser source noise

We noted [Equation (2.13)] that for an analog system, a time-varying signal $f(t)$ is used to directly modulate the optical source about a bias point I_B (see Figures 1.22 and 2.14). With no input signal present, the optical power is P_t. When the input signal is applied, the optical output power $P(t)$ is given by

$$P(t) = P_t[1 + mf(t)] \tag{2.47}$$

In this expression, m is the modulation depth defined by

$$m = \frac{\Delta I}{I_B'} \tag{2.48}$$

where

$I_B' = I_B - I_T$

I_T = laser threshold current

ΔI = RF current variation about the bias point

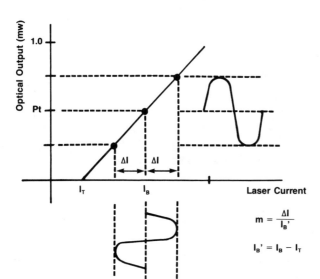

Figure 2.14 Laser light intensity curve. (From *RF Design,* April 1987, with permission.)

As noted previously, in order to prevent distortions in the output signal, the modulation must be confined to the linear region of the curve. Furthermore, if $\Delta I > I_B'$, the lower portion of the signal is clipped, and severe distortion occurs. For analog applications, typical values for m range from 0.25 to 0.50. A higher modulation index will provide a higher RF CNR, since the received RF carrier is proportional to m. Thus, there exists a direct trade-off between system distortion and noise performance due to contributions by the optical transmitter.

We noted in Section 2.1.1 that inherent laser source noise places an upper bound on the maximum achievable carrier-to-noise ratio obtainable from a laser transmitter. When the laser is biased above threshold, minute fluctuations in optical emission are exhibited. This noise phenomenon is referred to as *relative intensity noise* (RIN) and is neither thermal nor strictly shot noise in nature. It is the response of the laser to modulation by intrinsic shot noise. Shot noise, itself, results from the granular nature of light and electricity. Typical values of laser CNR due to RIN are −120 to −140 dB/Hz.

2.4.2 Fiber/Receiver Carrier-To-Noise Contribution

We introduced in Chapter 1 the concept of an optical loss budget. This is the conventional approach used for determining the maximum optical path loss. The optical power transmitted is compared with the minimum optical power at the receiver to provide the required CNR out of the receiver. The difference between transmitted and received power (total path loss) is the allowable optical loss budget. A typical loss budget is given in Table 2.2.

TABLE 2.2   OPTICAL LOSS BUDGET

Power launched (Tx)	−3.0 dBm
Fiber loss	20.0 dB
Splice loss	2.0 dB
Connector loss	1.0 dB
Received optical power (Rx)	−26.0 dBm
Receiver sensitivity (Rx) (minute power allowable to provide required C/N)	−30.0 dBm
System margin	4.0 dB

Generally speaking, the systems engineer is interested in how the CNR performance of the system changes as a function of the optical input power. The answer to this depends on whether the system noise performance is receiver-limited, quantum-noise-limited, or laser-source-noise-limited. The complete expression for the CNR present at the output of an optical receiver employing a PIN diode is given by

$$\frac{C}{N} = \frac{(1/2N^2)(mR_\lambda P_r)^2}{\underbrace{(\text{RIN } R_\lambda^2 P_r^2 B)}_{\text{source}} + \underbrace{2e(R_\lambda P_t + I_d)B}_{\text{quantum}} + \underbrace{(4kTB/R_{eq})F}_{\text{receiver}}} \quad (2.49)$$

where

R_λ = diode responsivity

m = modulation depth

P_r = average optical power received

e = electron charge

k = Boltzmann's constant

I_d = diode dark current

B = receiver bandwidth

T = temperature, °K

R_{eq} = equivalent resistance of photodiode load and amplifier

F = preamplifier noise figure

RIN = source relative intensity noise

N = number of FDM channels

P_t = transmitted optical power

When the optical power incident on the photodiode is small, the receiver circuit noise term dominates the system noise, and Equation (2.49) reduces to

$$\frac{C}{N} = \frac{(1/2N^2)(mR_\lambda P_r)^2}{(4kTB/R_{eq})F} \quad (2.50)$$

In this expression, the CNR is directly proportional to the square of the average received optical power. Thus, for each 1-dB change in optical power received, the CNR ratio will change by 2 dB. For larger optical signals incident on the photodiode, the quantum noise associated with the signal detection process dominates, and Equation (2.49) reduces to

$$\frac{C}{N} = \frac{(1/2N^2)(mR_\lambda P_r)^2}{2eB} \quad (2.51)$$

This expression assumes that the diode dark current i_d is negligible. Since the CNR for this case is independent of circuit noise, Equation (2.51) represents the fundamental or quantum limit for analog receiver sensitivity. In this optical power range, the CNR will change 1 dB for each 1 dB in received optical power.

As a final consideration, for very high optical power levels, the CNR may be limited by the laser source, and Equation (2.49) becomes

$$\frac{C}{N} = \frac{(1/2N^2)(m^2)}{(RIN)(B)} \quad (2.52)$$

That is, the CNR is constant at the maximum obtainable from the laser transmitter. Figure 2.15 illustrates an example of CNR obtainable at the receiver output as a function of optical input power when the various noise sources are present.

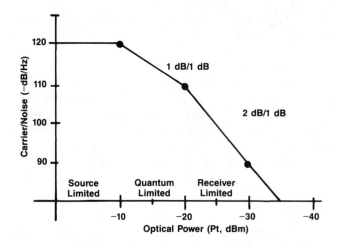

Figure 2.15 Carrier-to-noise ratio versus received optical power. (From *RF Design,* April 1987, with permission.)

2.4.3 Bandwidth Limitations

The modulation response of a laser diode is given in Figure 1.21. Laser transmitters have available modulation bandwidths of at least several GHz. Generally speaking, the transmitter circuitry will limit the upper bandwidth rather than the semiconductor laser diode. Linearity constraints, however, will tend to limit full utilization of the available bandwidth in analog FDM systems. Single-mode 1300-nm fibers are available with bandwidths exceeding 10 GHz/km. The electrical 3-dB bandwidth for a single–mode fiber is given by

$$f_{-3\text{ dB}} = \frac{0.35}{(M)\,(\Delta\lambda)\,(L)} \tag{2.53}$$

where

M = material dispersion (ps/nm × km)

$\Delta\lambda$ = optical spectral width

L = fiber length (km)

Example 2.5

A single-mode 1300-nm fiber is available with material dispersion equal to 0.035 (ps/nm × km). Calculate the 3-dB bandwidth for a 10-km length. Assume an optical spectral width (single-mode laser) of 1 nm.

Solution From the statement of the problem

$$M = 3.5 \text{ (ps/nm × km)}$$

$$\Delta\lambda = 1 \text{ nm}$$

$$f_{-3\text{ dB}} = \frac{0.35}{(3.5 \text{ ps/nm × km})(1 \text{ nm})(10 \text{ km})} = \frac{0.35}{35 \times 10^{-12}}$$

$$= 10 \text{ GHz}$$

High-sensitivity receivers require that the bandwidth be limited. Note from the expression for CNR [Equation (2.49)] that all of the noise terms are directly proportional to bandwidth. In an optical system design the bandwidth is intentionally limited by the optical receiver to only that required to transmit the information. Note also from the expression that high-sensitivity receivers require high input resistance. However, as has been noted previously, it is difficult to maintain high bandwidth at the same time. This is due to real and parasitic capacitances present in the optical receiver circuitry. Typically, very wide-band (>1 GHz) receivers utilize 50-ohm photodiode, preamp, and postamp stages. This results in lower receiver sensitivity due to the 50-ohm thermal noise source at the preamp input.

2.4.4 Distortion

The limiting distortions in an analog FDM system are second- and third-order intermodulation products. Intermodulation product distortion (IMD) occurs when two large unwanted signals beat together (intermodulate) in a nonlinear device to produce a product at the unwanted frequency. In a fiber optic link, the laser diode is normally the distortion-limiting component. Second-order IMD products result in frequencies of $(f_1 \pm f_2)$, where f_1 and f_2 are the two first-order unwanted frequencies. Third-order IMD products are defined as $(f_1 \pm f_2 \pm f_3)$ products. Typical values of distortion for a single-mode laser operating at a 50 percent modulation depth are

- Second order: 30–45 dB
- Third order: 45–60 dB

The spread in values results from the fact that not all lasers have good linearity. In fact, some lasers may even have abrupt discontinuities in their light intensity curves that disqualify them completely for use in analog systems.

To date, circuit linearization techniques for lasers have not proved successful. Work is continuing in this area, however, particularly for AM modulation. At this time, careful specification criteria and selection of lasers for linear analog performance is the best approach. Since signal levels are low, the linearity of fiber optic receivers is quite good. PIN photodiodes have good linearity over several orders of magnitude. When very high optical powers are present at the receiver input, it may be necessary to utilize an optical or an electrical automatic gain control (AGC). If the high level is a permanent condition, a fixed optical attenuator may be used.

If the systems designer has prior knowledge of the distortion performance of a fiber optic link, it is possible to take steps to minimize the distortion. For example, since second orders are the strongest distortions, a frequency transmission spectrum can be selected that eliminates second orders from falling in desired channels (see Figure 2.16).

Without the second-order limitation, third-order products will be the limiting distortion mechanism. As the number of channels increases above three, then *beat stacking* will occur (Figure 2.17). That is, several individual beats will exist at or

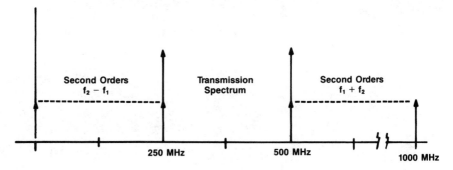

Figure 2.16 Spectrum plan to avoid second orders. (From *RF Design*, April 1987, with permission.)

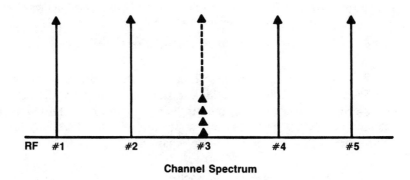

Figure 2.17 Beat stacking (for 5-channel system). (From *RF Design*, April 1987, with permission.)

near the same frequency. This is commonly referred to as composite triple beats (CTB). The result of beat stacking tends to be additive on a power basis. This assumes that the individual carriers are not phase-locked. The cumulative effect of stacking can be estimated by using the following expression:

$$\text{CTB} = -d3 + 10 \log (N) \qquad (2.54)$$

where

- $-d3$ = distortion level of a single third-order intermode (three-tone) product
- N = number of stacked beats (equal levels)

For a symmetrically spaced spectrum, the largest number of beats will fall in the center spectrum. An example will illustrate this.

Example 2.6

In Figure 2.17, the FDM channels are spaced 50 MHz apart with the lowest channel at 150 MHz. Find the worst-case increase in distortion in the center channel over that contributed by a single third-order distortion.

Solution

$$f_3 = f_1 + 2(50)$$
$$= 150 + 2(50) = 250 \text{ MHz}$$

In this particular case, the desired frequency is f_3 (250 MHz). Any third-order beat frequencies are produced by f_1, f_2, f_4, and f_5. An examination of these frequencies results in the following:

$$f_1 - f_2 + f_4 = 150 - 200 + 300 = 250 \text{ MHz}$$
$$f_2 - f_4 + f_5 = 200 - 300 + 350 = 250 \text{ MHz}$$
$$f_4 - f_2 + f_1 = 300 - 200 + 150 = 250 \text{ MHz}$$
$$f_5 - f_4 + f_2 = 350 - 300 + 200 = 250 \text{ MHz}$$

Thus, there are four (three-tone) third-order beats stacked in the center of the channel (f_3) spectrum. The increase in distortion is

$$10 \log (N) = 10 \log (4) = +6 \text{ dB}$$

Two-tone, third-order distortions will also be present with multiple carriers. However, they will have individual distortions that are 6 dB below a three-tone third-order product and thus tend to be a negligible contribution as the number of channels increases. These extrapolations are useful, since many linearity tests are performed with a limited number of channels.

2.4.5 Alternative Design Approaches

The preceding discussion does not represent the only design approach for FM systems. A low-cost yet high-performance, color, composite video fiber optic link for short-wave applications can easily be built by using readily available off-the-shelf integrated circuits (Figure 2.18).

Starting at the transmitter end, the system begins with an NE592 differential video amplifier. The amplifier receives the composite video signal and then differentially drives the voltage-controlled oscillator (VCO) of an NE564 phase-locked loop (PLL). This is done through the output pins of the PLL phase comparator. The VCO is driven directly to avoid the input limiter and phase detector of the NE564. This method of operation opens up a number of applications for the NE564 that were previously impossible. The loop, in this case, is configured as a frequency modulator with a center of 30 MHz and a deviation of ±10 MHz. From here, the modulated signal is fed to an NE522 high-speed comparator with an open collector output. The comparator boosts the signal in order to drive a high-power aluminum gallium arsenide infrared (180 μm) LED that has a typical rise time of 3 ns.

The composite video is essentially sent over the fiber and is received by an AlGaAs PIN photodiode. The light is converted to a current by the diode and is amplified and changed to a voltage by the NE5539 op-amp in a transimpedance configuration. The very high-speed response (600 V/μs) and wide bandwidth 350-MHz unity gain) makes this device ideally suited for high-performance optical links.

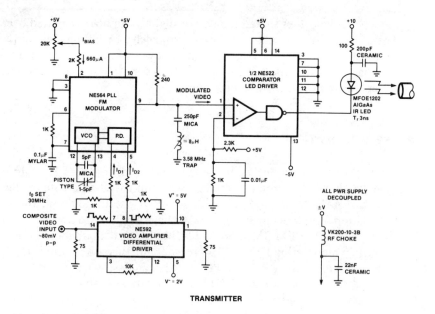

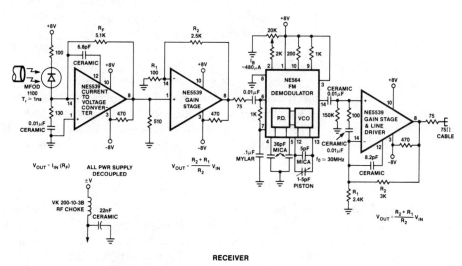

Figure 2.18 Fiber optic video system. (Reprinted with permission from Signetics Company, a division of North American Phillips Corporation, AN 146, OUT, 1984.)

Compensation components and their values are also shown in Figure 2.18 to make the NE5539 a voltage gain stage and is optional depending upon the attenuation in the fiber or upon its length. Immediately following is another NE564 PLL set up as an FM demodulator that is AC-coupled to the last NE5539 op-amp. This third NE5539 acts as an amplifier and buffer that is capable of driving a 75-ohm cable to a video monitor.

Sec. 2.4 Design of Frequency Division Multiplexed Fiber Optic Links

Note that other signals (analog or digital) can be transmitted over this circuit. These signals can be sent over lower frequencies. Additional circuitry can be used to multiplex the signals after the first-phase locked loop in the transmitter, and bandpass filters can be utilized before the second-phase locked loop in the receiver to separate and then demodulate them.

2.5 PULSE MODULATION

In the preceding section, we looked at techniques in which RF subcarriers can be modulated and the summed outputs used to intensity-modulate the light. Thus, we can have AM-IM, FM-IM, PM-IM, and so on. In addition, the familiar pulse modulation techniques are available. We have available to us such schemes as PAM-IM, PWM-IM, PPM-IM, and PFM-IM. In particular, there is interest in pulse frequency modulation (PFM). The reasons for this are twofold: (1) Results in the literature indicate that PFM is the yardstick against which the other pulse modulation techniques are measured; (2) there is ready availability of components needed to implement the system.

As noted in Chapter 1, the rise time plays a prominent part in pulse-type systems (see Section 1.3.8). Before we discuss pulse-type systems, we consider the end-to-end system rise time of a fiber optic system.

2.5.1 System Rise Time

A consideration for system bandwidth must be the overall system rise time. The source and detector rise times, for example, can distort the shape of a signal (pulse). Typically, these range from 5 to 15 ns for LEDs and 0.1 to 2 ns for ILDs. At the receiving end, PINs and APDs have rise times between ±4 ns. The total rise time may be expressed as

$$T_{system} = 1.1 \sqrt{T_{source}^2 + T_{modal}^2 + T_{material}^2 + T_{detector}^2} \qquad (2.55)$$

where T equals the 10 percent to 90 percent rise time. The weighting factor of 1.1 results from a comparison of the rise-time edge with the overall pulse dispersion. The -3-dB bandwidth, and therefore the maximum bit rate, is usually defined in terms of T_{system} and the rise time of the RC circuit (Figure 2.19).

With a step-input voltage amplitude V, the output voltage, $V_{out}(t)$, is given by

$$V_{out}(t) = V(1 - e^{-t/RC}) \qquad (2.56)$$

and the 10 percent to 90 percent rise time t_r is

$$t_r = 2.2RC \qquad (2.57)$$

The transfer function [Figure 2.18(b)] is readily determined to be

$$H(jw) = \frac{1}{1 + j\omega RC} \qquad (2.58)$$

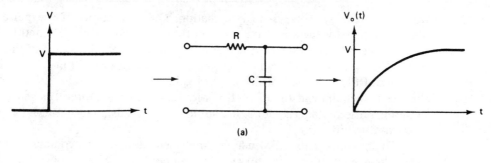

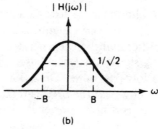

Figure 2.19 (a) Response of a low-pass RC filter circuit to a voltage step input. (b) Transfer function $H(\omega)$ for the circuit in (a).

From this expression we have the familiar result for the -3-dB bandwidth; that is

$$\text{BW} = \frac{1}{2\pi RC} \qquad (2.59)$$

Combining Equations (2.57) and (2.59) results in

$$t_r = \frac{2.2}{2\pi} = \frac{0.35}{\text{BW}} \qquad (2.60)$$

where $t_r = T_{\text{system}}$ for a fiber optic link.

Obviously, different results are obtained for different filters. For fiber optic systems, the constant 0.35 is a conservative number and is often used. In order to retain the shape of the pulse at the filter output with reasonable fidelity, the -3-dB bandwidth must be at least large enough to satisfy the condition $\text{BW} \cdot \tau = 1$, where τ is the pulse duration. Thus from Equation (2.60) we have

$$T_{\text{system}} = t_r = 0.35\tau \qquad (2.61)$$

We will apply these results in the next section and again in Chapter 3.

2.5.2 Pulse Frequency Modulation

All of the pulse analog techniques employ pulse modulation in the electronics prior to intensity modulation of the optical source. PWM–IM is affected by nonlinearities and is usually discounted. PWM–IM is also inefficient, since a large part of the transmitted energy conveys no information. Only variations of the pulse width about a nominal value are of interest. On the other hand, PPM-IM and PFM-IM offer distinct advantages, since the modulation affects only the timing of the pulses, which allows the transmission of very narrow pulses. PFM-IM provides a greater SNR than PPM-IM, since wideband FM gain can also be obtained. Furthermore, the terminal equipment for PFM-IM is less complex; therefore, it is generally the preferred pulse analog technique. In addition to these advantages, PFM-IM permits frequency multiplexing.

The modulator for PFM may be implemented by using a commercially available voltage-controlled oscillator (VCO). Demodulation is accomplished by using a pulse regenerator followed by a low-pass filter. A block diagram for a PFM-IM optical fiber system is shown in Figure 2.20. The received pulses are detected in a wideband preamplifier and then regenerated with a limiter and a "one-shot." The original modulating signal exists as a baseband component, and recovery is accomplished by using a low-pass filter. Detection can also be accomplished without the pulse regeneration; however, the SNR is reduced. The SNR in terms of the peak-to-peak signal power to rms noise power is given by

$$\left(\frac{S}{N}\right)_{p\text{-}p} = \frac{12(T_D \Delta f G_0 R_\lambda P_{so})^2}{(2\pi t_r B)^2 \, \overline{i_{\text{eff}}^2}} \tag{2.62}$$

where

$T_D = \dfrac{1}{f_0}$ = normal pulse period

Δf = frequency deviation

G_0 = photodiode (avalanche) multiplication factor

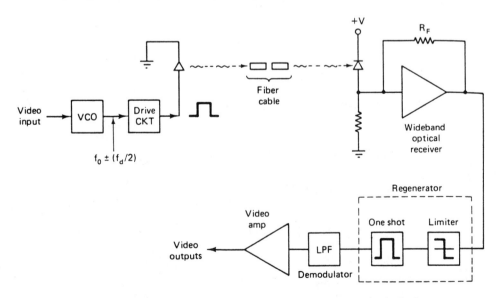

Figure 2.20 PFM-IM optical fiber system employing regenerative baseband recovery.

P_{so} = peak receiver optical power

t_r = pulse rise time at the regenerator circuit input

B = baseband noise bandwidth, Hz

i^2_{eff} = receiver mean square noise current

R_λ = responsivity

Example 2.7

An optical fiber system is to be designed to operate with repeater spacing at 8 km. Components available have the following specifications:

Fiber: intermodal	5 ns/km
Pulse spreading: intramodal	1 ns/km
Pin diode	7 ns
LED	7 ns

For the receiver in Figure 2.20, calculate the peak-to-peak signal-to-rms-noise ratio. Assume that an APD photodiode is used in the receiver and the received optical power is -40 dBM. The following system parameters apply:

$$\text{Rest pulse rate} = 15 \text{ MHz}$$
$$\text{Peak frequency deviation} = 1.5 \text{ MHz}$$
$$\text{APD responsivity} = 0.7$$
$$\text{Baseband noise bandwidth} = 5 \text{ MHz}$$
$$\text{Receiver mean square noise current} = (0.45 \times 10^{-9})^2 A^2$$

Solution The pulse rise time at the regenerator circuit should equal the system rise time for an optimum system. The total system rise time is

$$T_{system} = 1.1[7^2 + (8 \times 5)^2 + (8 \times 1)^2 + 7^2]^{1/2}$$

$$= 46.16 \text{ ns}$$

$$t_r = T_{system} = 46.17 \text{ ns}; \qquad -40 \text{ dBm} \rightarrow 10^{-7} \text{ W}$$

$$T_D = \frac{1}{f_0} = \frac{1}{15} \times 10^{-6} = 6.67 \times 10^{-8} \text{ s}$$

$$\left(\frac{S}{N}\right)_{p\text{-}p} = \frac{12[6.67 \times 10^{-8})(1.5 \times 10^6)(60)(0.7)(10^{-7})]^2}{[(2)(46.17 \times 10^{-9})(5 \times 10^6)]^2[0.45 \times 10^{-9}]^2}$$

$$= 4.98 \times 10^{-8} \rightarrow 73 \text{ dB}$$

2.5.3 Multichannel PFM Video (FDM)

We noted in Section 2.4 that multichannel signals can be applied as intensity modulation to the optical transmitter. This applies also to PFM signals. Only the relative timing of the zero crossing information is used in reconstruction of the multiple channels at the receiver, thus reducing the effects of system nonlinearities. Figure 2.21 illustrates representative architecture which incorporates PFM channels that are frequency-multiplexed onto a single-mode fiber. The output of each PFM modulator is a unique VHF PFM signal that is minimally bandlimited to control interfer-

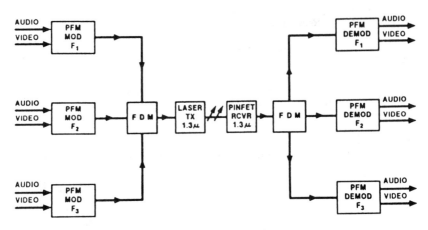

Figure 2.21 Electronically multiplexed PFM full-motion video fiber optic system.

ence with adjacent laser-modulating PFM carriers at different VHF frequencies. Increasing the amount of per-channel filtering causes the pulse rise/fall time to increase. The performance deteriorates and eventually approaches that of the FM case. The three PFM carriers are summed at the optical transmitter input, and this composite electrical signal is used to intensity-modulate the laser. The design of the laser transmitter was discussed in Section 2.4 (see Figure 2.12).

At the receiving end of the link, light is coupled from the fiber into a PIN FET or linearized APD receiver (see Figure 2.13). The receiver circuitry provides additional amplification and gain control to maintain the received multiplexed electrical output signal at the required level. The composite signal is demultiplexed (filtered) into the respective VHF PFM channels. Regenerative receivers then recover the baseband signal (see Figure 2.20).

For multichannel fiber video applications, the frequencies of the individual VHF PFM modulating carriers are selected to minimize the overall system intermodulation products. This technique was discussed in Section 2.4.4. However, because laser intensity modulation is employed to transport the multiplexed signal, the laser must be carefully selected for high linearity to prevent the generation of excessive intermodulation interference. Bandwidth limiting of individual PFM signals at the transmitter and receiver can reduce interference and can also reduce the total optical modulation bandwidth required. However, this occurs at the expense of output signal-to-noise ratio (rise/fall times are increased).

For long-haul transmission of video, digital systems are the systems of choice. This is because digital systems can be made distance-independent by using regenerating repeaters. In both the PFM and digital case, the number of channels can be increased by using wavelength division multiplexing (WDM). This can be accomplished without increasing the end-to-end fiber bandwidth requirements. We may wavelength multiplex, for example, at 1.2, 1.3, and 1.55 microns. Care must be taken in wavelength separation to minimize optical cross talk.

In today's video transmission networks over fibers, PFM and digital techniques often must interface. Transmitting digital information over fibers is discussed in Chapter 3.

REFERENCES

2.1. URY, ISRAEL. 1985. Optical communications. *Microwave Journal*, 24–35.
2.2. LAU, K. Y., et al. 1984. Wideband laser diodes spark new designs. *Microwaves and RF*, 109–116.
2.3. KOSCINSKI, J. 1987. Transmission of analog FDM signals on fiber optic links. *RF Design*, 31–41.

PROBLEMS

2.1. (a) If a 5-mW signal with $\lambda = 1.55$ μm is coupled into a single-mode optical fiber, what is the power 100 km away if the attenuation is 0.2 dB/km?
 (b) What current will this produce in a photodetector with a quantum efficiency of 60 percent?

2.2. A baseband signal with unit power and bandwidth 1 MHz is RF subcarrier-modulated with the subcarrier intensity-modulated onto an optical carrier at 10^{14} Hz. The optical system is quantum-limited. A subcarrier SNR of 20 dB is required for subcarrier demodulation.
 (a) Calculate the required received optical power to ensure proper subcarrier operation.
 (b) What is the baseband SNR after subcarrier demodulation? Assume AM/IM modulation.
 (c) What is the baseband SNR after subcarrier demodulation if FM/IM is used with a subcarrier deviation of 10 MHz?
 (d) What subcarrier bandwidths are needed in (b) and (c)?
 (e) If the ratio of background to received signal power is 0.5 (no longer quantum-limited), how much more optical power is needed in (a)?

2.3. Determine the wavelength in nanometers and angstroms (1 angstrom = 10^{-10} m) for the following light frequencies:
 (a) 3.45×10^{14} Hz
 (b) 3.62×10^{14} Hz
 (c) 3.21×10^{14} Hz

2.4. Prove Equation (2.33).

2.5. Repeat Example 2.3 with an avalanche photodetector gain of 40.

2.6. Refer to Figure 2.17. The FDM channels are spaced 40 MHz apart, with the lowest channel at 100 MHz.
 (a) Find the worst increase in distortion in channel 2 over that contributed by a single channel.
 (b) Do there exist any second-order IMD products that will interfere with any of the channels?

2.7. Plot Equation (2.49) (CNR in dB/Hz versus P_r) for the following parameters:

$P_t = -3$ dBm
$N = 18$ channels
BW = 50 MHz/channel
Responsivity = 0.8
Modulation depth = 3%/channel
$R_{eq} = 50$
Dark current = 2 amps
$T = 27°K$
Receiver noise figure = 3 dB
Relative intensity noise = -123 dBm

Compare your results to Figure 2.15.

2.8. An analog fiber optic communication system requires an SNR of 40 dB at the detector with a postdetection bandwidth of 30 MHz. Calculate the minimum optical power required at the detector if it is operating at a wavelength of 0.9 μm with a quantum efficiency of 70 percent. State any assumptions made.

2.9. A silicon PIN photodiode has a quantum efficiency of 65 percent and operates at a wavelength of 0.8 μm. Find:
  (a) the mean photocurrent when the detector is illuminated at a wavelength of 0.8 μm with 5 μW of optical power.
  (b) the rms quantum noise current in a postdetection bandwidth of 20 MHz.
  (c) the SNR in dB, when the mean photocurrent is the signal.

2.10. The photodiode in Problem 2.9 has a capacitance of 8 pF. Calculate:
  (a) the minimum load resistance corresponding to a postdetection bandwidth of 20 MHz.
  (b) the rms thermal noise current in the above resistance at a temperature of 25°C.
  (c) the SNR in decibels resulting from the illumination in Problem 2.9 when the device dark current is 1 nA.

2.11. The photodiode in Problems 2.9 and 2.10 is used in a receiver, where it drives an amplifier with a noise figure of 3 dB and an input capacitance of 8 pF. Calculate:
  (a) The maximum amplifier input resistance to maintain a postdetection bandwidth of 20 MHz without equalization.
  (b) The minimum incident optical power required to give an SNR of 50 dB.

2.12. A germaninum photodiode is used in an optical receiver. The diode operates at a wavelength of 1.55 μm and has a dark current of 400 nA at the operating temperature. When the incident optical power is 10^{-6} W and the responsivity of the diode is 0.6 A/W, shot noise predominates in the receiver. Determine the SNR at the receiver when the postdetection bandwidth is 50 MHz.

3

Digital Fiber Optic System Design

3.0 INTRODUCTION

After a decade of exponential growth in performance, lightwave technology is still advancing rapidly. This advancement has been so rapid that today an optical fiber can carry higher data rates over greater distances than has ever been possible with other transmission media. In this chapter, we wish to apply the basic principles discussed previously to the design of optical links for transmitting digital data.

Consider the digital link shown in Figure 3.1, consisting of a transmitter, a fiber transmission medium, and a receiver. The transmitter converts the incoming binary data to on-off light pulses, which are launched into the fiber. At the receiver, the modulated optical signal is detected (demodulated), and the data are recovered as electrical signals (direct detection). Prior to 1970, a link such as this could not be used to transmit over distances greater than a few hundred meters. In 1970, Corning Glass succeeded in producing a fiber with an attenuation of approximately 20 dB/km. This made transmission of a few kilometers commercially feasible. This event, coupled with the simultaneous development of semiconductor light sources, resulted in a worldwide explosion of lightwave research and development, which continues today.

The design of fiber optic systems proceeds in three steps: selecting the wavelength, selecting a component set, and using the loss and time budgets to evaluate

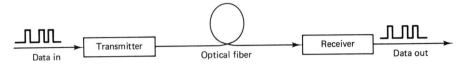

Figure 3.1 Simple lightwave communications link. (From P. S. Henry, "Introduction to Lightwave Transmission," *IEEE Communications Magazine,* May 1985. © 1985 IEEE.)

the design. Wavelength selection is based on the data rate and distance for the system. Low-speed systems and short distances allow the use of long wavelengths, while high-performance systems require shorter wavelengths. Component selection is also based on system performance requirements. For example, LEDs and step-index fibers are suitable only for very short distances, while lasers with single-mode dispersion-shifted fiber can transmit data with rates of gigabits per second over hundreds of kilometers.

We use loss budgets to determine if the received power is greater than the sensitivity of the receiver. Too little power results in an increase in the bit error rate. The time is used to verify if the overall system rise time and, therefore, bandwidth are adequate for the specified data rate and/or encoded data. For example, we shall see that Manchester encoding requires twice the bandwidth of nonreturn-to-zero (NRZ) coding. If a system fails to meet system specifications, the loss and time budgets can also be used to determine the best solution.

After the system is designed and constructed, tests must be performed to verify performance. Specifically, three system checks can be used to determine system performance. These are

- Measure the DC output power of the source and fiber, using a power meter.
- Measure power at all points along the fiber, using an optical time domain reflectometer (OTDR).
- Measure the digital data transmission quality, using a bit error rate (BER) tester.

Before we consider the design of digital fiber links, we wish to look at a number of system-level parameters that affect system performance.

3.1 DIGITAL SYSTEM PERFORMANCE PARAMETERS

It is convenient to consider as a measure of fiber link performance the product of the bit rate B times the length L of the fiber. This simple figure of merit, $B \cdot L$, is consistent with the intuitive notion that both bit rate and distance are important contributors to the overall performance measure of a system. Consider the system in Figure 3.2, which illustrates a lightwave system consisting of several repeatered lines in parallel. Let the objective be to transmit a total bit rate of $B_T > B$ over a

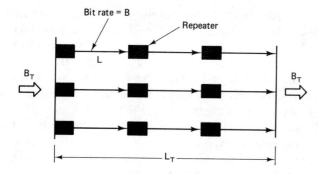

Figure 3.2 Repeatered lightwave system. The total number of repeaters is inversely proportional to B•L for each individual link.

distance $L_T > L$, where B and L are the bit rate and fiber length of each individual repeater section. The total number of repeaters N required for the system is

$$N = \frac{B_T \cdot L_T}{B \cdot L} \tag{3.1}$$

Thus, to minimize the number of repeaters in a section, we desire to use technology that provides the greatest $B \cdot L$ product. Referring to Equation (3.1), we see that both B and L are equally important in determining the "box count" in Figure 3.2. However, we shall see that with current technology, the most attractive route for maximizing $B \cdot L$ is to operate at the highest possible bit rate, rather than strive for ultralong repeater sections.

With the exception of very short links, such as may be appropriate for intrabuilding networks, the most important consideration in lightwave system performance is fiber loss. As an optical pulse propagates along the fiber, it is attenuated exponentially. At a distance L km from the transmitter, the signal power is given by

$$P(l) = P_T 10^{-Al/10} \tag{3.2}$$

where P_T is the transmitted power and A is the fiber attenuation expressed in dB/km. The attenuation of high-quality fibers is shown in Figure 3.3. Two mechanisms

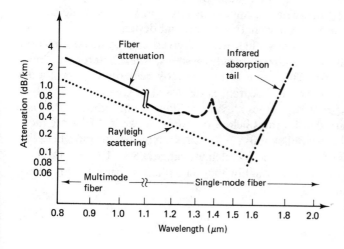

Figure 3.3 Attenuation of high-quality silica fibers. The dashed peaks correspond to OH absorption, which can (in principle) be eliminated. In general, multimode fibers tend to have higher loss than single-mode, because of increased scattering from dopants in the fiber core. (From P. S. Henry, "Introduction to Lightwave Transmission," *IEEE Communications Magazine,* May 1985. © 1985 IEEE.)

account for most of the observed loss. At short wavelengths, Rayleigh scattering from inevitable density fluctuations "frozen" into the fiber during manufacture leads to attenuation inversely proportional to the fourth power of the wavelength. At longer wavelengths, the attenuation rises rapidly because of absorption from the tails of the infrared resonances in silica and other fiber constituents. These two processes define a low-loss window with a minimum attenuation of ≈ 0.2 dB. As noted in Chapter 1, this occurs at a wavelength near 1.5 μm.

The link shown in Figure 3.1 can operate at an acceptably low bit-error rate only if the optical power at the receiver exceeds a minimum power level called the *receiver sensitivity*. The exponential attenuation given by Equation (3.2) implies a maximum loss-limited transmission distance, which can be expressed as

$$L_{max} = \frac{10}{A} \log_{10} \frac{P_T}{P_R} \qquad (3.3)$$

where, as previously noted, P_T is the transmitted power. We can draw two conclusions from this equation: (1) L_{max} is relatively sensitive to changes in the fiber attenuation A, and (2) L_{max} is weakly dependent on the transmitter power and receiver sensitivity. In a typical fiber system, for example, $P_T/P_R \approx 10^4$. Thus, increasing the transmitter power by an order of magnitude increases L_{max} by only about 20 percent. This is in contrast to radio systems, where signal attenuation obeys an inverse square law, and a tenfold increase in transmitter power more than triples the range. The desire to increase L_{max} by reducing fiber attenuation has driven the gradual evolution of operating wavelength from 0.8 μm to 1.55 μm. This also forced a change from gallium-arsenide-based optical sources to an entirely new long-wavelength technology based on indium phosphide.

We can use Equation (3.3) to estimate values for L_{max}. In order to accomplish this, we need representative values for P_T and P_R. We know at this point that the optical sources are almost invariably LEDs or laser diodes. Furthermore, the laser is preferable to the LED because of its superior modulation-speed capability (> 4 Gbits/s) and better optical coupling efficiency into single-mode fibers. Generally speaking, most lasers used in lightwave applications deliver roughly the same amount of power into the fiber. That is, approximately 1 mW. This is the case in spite of significant differences in wavelength and internal design.

We noted in Chapter 2 that today's lightwave receivers employ direct direction of the incoming signal. This is in contrast to radio systems, where heterodyne down conversion is performed first. This same technique (coherent optical detection), presently the subject of considerable research, is discussed in Chapter 7. The two choices of photodetectors are PIN photodiodes and avalanche photodiodes (APDs). Generally, PINs are preferred at bit rates below 100 Mbits/s, while APDs are most advantageous for high bit rate applications ($\geq$ 1 Gbit/s). Now for bit rates between 10 Mbits/s and 2 Gbits/s, and over the wavelength range 0.8 to 1.6 μm, the best lightwave receivers requires a flux of roughly 300–1000 photons per bit. This corresponds to a receiver sensitivity P_R of

$$P_R \sim 2 \times 10^{-13} B \text{ mW} \qquad (3.4)$$

where B is the bit rate. Curves of L_{max} are compound and are illustrated in Figure 3.4 with Equations (3.3) and (3.4). The curve for single-mode fiber should not be taken too literally, because beyond 4 Gbits/s there has been no experimental verification of Equation (3.4). Note that the nearly horizontal shape of these curves demonstrates that it is possible to increase $B \cdot L$ by making large increases in the bit rate B with only a small penalty in the length L. This is true until the bit rate becomes so high that delay-spread due to fiber dispersion becomes significant.

Efforts to improve receiver sensitivity continue. It is shown in the literature that the probability of error for a quantum-limited detector is given by [see Equation (2.49)]

$$P_e = \exp\left[-\frac{\eta P_s}{hfB}\right] \tag{3.5}$$

where

P_s = incident optical power

η = quantum efficiency of the photodiode

hf = photon energy = $\dfrac{2 \times 10^{-19}}{(\lambda/\mu m)}$

B = bit rate (per second)

For a probability of error $\leq 10^{-9}$, the number of photons required per bit is

$$\eta P_s \geq 21 \text{ photons per bit} \tag{3.6}$$

We can appreciate the relative effects of circuit and source noise. The power required per bit is then

$$\eta P_s = 21\, hfB = 4.2 \times 10^{-12} \left[\frac{B/\text{Mbits/s}}{\lambda/\mu m}\right] \text{W} \tag{3.7}$$

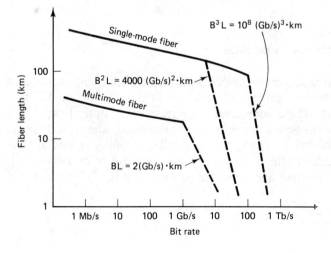

Figure 3.4 Lightwave performance limits. The solid curves show the distance limit imposed by fiber loss (2 dB/km for multimode at 0.8 μm wavelength; 0.2 dB/km for single-mode at 1.55 μm wavelength). The dashed curves show dispersion limitations associated with different dispersive mechanisms described in the text. The solid circle (•) represents the highest reported B•L performance: 2 Gb/s over 130 km, or B•L = 260 Gb/s • km [11]. (From P. S. Henry, "Introduction to Lightwave Transmission," *IEEE Communications Magazine,* May 1985. © 1985 IEEE.)

The resultant current is illustrated in Figure 3.5 and may be calculated by using Equation (2.3). That is,

$$i_s = \frac{\eta P_s q}{hf} \geq 21eB$$

$$= 3.4 \times 10^{-12} \left(\frac{B}{\text{Mbits/s}}\right) \text{ A} \qquad (3.8)$$

Example 3.1

Calculate the power and current required per bit for the following conditions:

$B = 4$ Gbits/s

$\eta = 60\%$

$\lambda = 1.5$ μm

Assume a quantum-limited detector. Compare with results obtained from Equation (3.4).

Solution

$$\eta P_s = 4.12 \times 10^{-12} \left(\frac{B/\text{Mbits/s}}{\lambda/\mu\text{m}}\right) \frac{1}{\eta}$$

$$= 4.12 \times 10^{-12} \left[\frac{4000}{(1.55)(0.6)}\right] = 1.772 \times 10^{-8} \text{ W}$$

$$L_s = 3.4 \times 10^{-12} \left(\frac{B}{\text{Mbits/s}}\right) = (3.4 \times 10^{-12})(4000)$$

$$= 1.36 \times 10^{-8} \text{ A}$$

Using Equation (3.4), we obtain

$$P_R \sim 2 \times 10^{-13} B = (2 \times 10^{-13})(4 \times 10^9) = 8 \times 10^{-7} \text{ W}$$

$$\frac{P_r}{P_s} = \frac{8 \times 10^{-7}}{1.77 \times 0.58} = 45.146$$

3.1.1 Dispersion Constraints on Bit Rate

We noted in Chapter 1 that dispersion arises because different components of a pulse travel at different speeds, and therefore arrive at the detector at different times. Refer to Figure 1.8. These components are the various optical rays. In a single-mode fiber they are the Fourier components of the pulse. The typical spread in propagation times of the different components is called the *delay-spread S*. Over a broad range of reasonable system models the degradation due to delay-spread is small, provided that S is less than half a symbol period. That is,

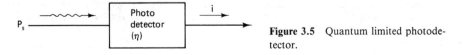

Figure 3.5 Quantum limited photodetector.

$$S < \frac{1}{2B} \tag{3.9}$$

In multimode fibers the dispersive mechanism is called *modal dispersion*. This was discussed in Chapter 1. Delay spread, in high-quality, graded-index multimode fibers, increases with propagation distance at a typical rate of 0.25 ns/km. Using Equation (3.9), we have

$$B \cdot L < 2 \text{ Gbits/s} \cdot \text{km} \tag{3.10}$$

This relation illustrates the dispersion constraint associated with multimode fibers and is the leftmost dashed line in Figure 3.4.

As noted previously, the dispersive effects in single-mode fibers are much smaller than in multimode fibers. In fact, modal dispersion is absent completely. The primary dispersive mechanism is material dispersion, which results from the frequency dependence of the refractive index of the fiber material. This is the same effect that causes a prism to disperse a beam of light into its constituent colors. The Fourier components of an optical pulse in a single-mode fiber propagate at different speeds. This results in a delay spread given by

$$S \sim D \cdot L \cdot W \tag{3.11}$$

where

D = absolute value of the linear dispersion, ns/km

L = fiber length, km

W = bandwidth of the pulse, Hz

The value of D for typical silica fibers is shown in Figure 3.6. Note that at a wavelength near 1.3 µm the linear dispersion goes to zero. At a wavelength of 1.55 µm, D is still very small (≈ 0.14 ps/km·GHz). A two-inch circular waveguide, by comparison, exhibits dispersion that is three orders of magnitude larger. The pulse bandwidth for an ideal transmitter (no intersymbol interference) is approximately equal to the bit rate. Using this fact, we find that Equations (3.9) and (3.11) yield the following dispersion constraint:

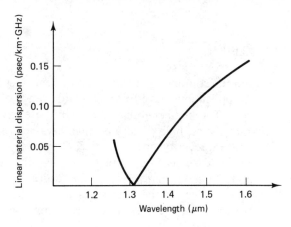

Figure 3.6 Linear material dispersion of silica fiber. Linear dispersion is zero near 1.3 µ wavelength, but at 1.55 µm it is large enough to cause significant impairments. (From P. S. Henry, "Introduction to Lightwave Transmission," *IEEE Communications Magazine*, May 1985. © 1985 IEEE.)

$$B^2L < 4000 \ (\text{Gbit/s})^2 \cdot \text{km} \qquad (3.12)$$

This equation is also plotted as a dashed line in Figure 3.4.

In practice, the delay spread given by Equation (3.12) represents a highly idealized situation in which the intrinsic spectral width of the unmodulated laser is small compared with the bit rate. This ideal, however, is being approached by nearly monochromatic single-frequency lasers. Most conventional lasers available today, however, exhibit spectral widths of several hundred GHz. That is, $W >> B$, which results in vastly increased pulse dispersion. Because of this problem, the earliest single-mode systems could not take advantage of the minimum in fiber loss at 1.55 µm. Instead, the systems had to operate in the higher loss wavelengths near 1.3 µm, where the dispersion effects were much reduced. Despite this, the 1.3-µm systems achieved impressive performance, which has only recently been surpassed by 1.55-µm systems.

A promising technique for achieving still higher $B \cdot L$ performance involves the use of a dispersion-shifted fiber. In this fiber the wavelength, $D = 0$, is shifted upward to coincide with the wavelength of minimum loss. Operation at this wavelength permits long-range transmission with only very small (second-order) dispersive effects. The delay spread, instead of being linear in B, becomes quadratic. This leads to a dispersion constraint given by

$$B^3L < 10^8 \ (\text{Gbit/s})^3 \cdot \text{km} \qquad (3.13)$$

This relation is shown in the far right dashed line of Figure 3.4. The maximum $B \cdot L$ product for this case is approximately 10^4 Gbit/s·km.

It is clear from Figure 3.4 that we are not yet at the limits of lightwave technology. Bit rates above 10 Gbits/s over spans of 100 km should be achievable. At this point, the primary technical challenge is not the fiber itself. Rather, it is the capabilities of the optical and electronic components at the ends of the fiber. One possibility may lie in directly modulating a single-frequency laser. If this proves not to be feasible, an unmodulated laser can be followed by an electro-optic modulator. These devices have been demonstrated to have modulation bandwidths greater than 10 GHz. At the receiver APDs are the choice. If the sensitivity deteriorates too rapidly above 4 Gbits/s, optical heterodyne techniques may prove to be the answer (see Chapter 7).

3.2 DESIGNING DIGITAL FIBER OPTIC LINKS

We see from the previous discussion the continuing evolution of optical fiber technology. Local optical transmission applications have fostered a wide range of new and different optimized components. The key benefits of fibers in these environments are superb attenuation characteristics, high data rate capability, electromagnetic immunity, ground loop elimination, security, and small size, along with expansion capability. This last fact is an increasingly important factor in using fibers. Fibers installed today will serve applications for many years in the future. Improvements in the technology will involve changes in the electronics only. Totally new installations will take maximum advantage of the latest in optical technology.

When a design is begun, system architecture concepts address such issues as tradeoffs between system transmission performance and complexity. For example, digital systems deal with parallel data. Thus, decisions concerning the degree of multiplexing versus the number of fiber channels are important. Slower parallel optical channels may save sufficient circuitry to justify their implementation. In other cases, data may occur in high data rate bursts. The choice is either a high-speed optical link, or buffering the data to provide a lower uniform data rate.

Once the data rate is established, methods for encoding the data are evaluated. Encoding for optical channels is considered in Chapter 4. For design purposes we will briefly introduce the subject at this time. Factors that influence the selection of a code are adding signal redundancy to enhance errorfree detection, enriching the data stream to ensure minimum data requirements, constraining the duty cycle of the signal to conform to the optical receiver, and including sufficient bit transmissions to allow clock recovery and synchronization. Some of the more common encoding formats are illustrated in Figure 3.7. In digital systems the encoding process consists of mapping from data bits to the signal elements illustrated. Choosing a particular code is significant in terms of the transmission system specification. Refer to Table 3.1. Note that the biphase code has two symbols per bit versus one symbol per bit for an NRZ code. Thus a 20-Mbit/s data rate signal encoded as biphase requires a link with twice the capacity in bits per second (40 Mbits/s) as does NRZ code. This effectively doubles the bandwidth requirements of the transmitter, receiver, and fiber. We shall see in Chapter 4, however, that the biphase code possesses a self-clocking capability, which the NRZ code does not. This is important in bit synchronizers, which must detect and reshape the received noisy bits.

Other codes, such as block codes, treat a number of bits as a set and encode the sets with a larger number of bits (Figure 3.8). In block codes, a block of *m*

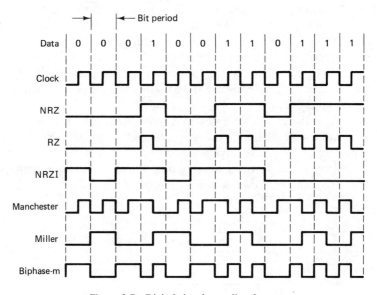

Figure 3.7 Digital signal encoding formats.

TABLE 3.1 DIGITAL CODE CHARACTERISTICS

Format	Symbols per Bit	Self-clocking	Duty Factor Range
NRZ	1	No	0–100
RZ	2	No	0–50
NRZI	1	No	0–100
Manchester (biphase-L)	2		50
Miller	1	Yes	33–67
Biphase-m (bifrequency)	2	Yes	50

information bits is followed by a group of r check bits. This results in a block of n code symbols. The group of r check bits serves to detect and, in some cases, correct errors. The n-bit block of output bits is called a *code word*. The code itself is referred to as an mB/nB code. Common types of group codes are the 1B/2B codes, such as the biphase code mentioned previously, 3B/4B, 4B/5B, and 8B/10B (see Figure 3.9). Note that the channel capacity increases only fractionally, at the cost of more complex encoding circuitry.

The specification for a system also determines the optional channel maximum and minimum distance required. For example, some applications may require longer length but with reduced throughput or accuracy requirements of other applications. An existing optical transmission system may provide the capability without additional redesign.

The error rate performance of the system must be factored in. Digital systems do make errors, and this is usually described as the bit error rate (BER) of the system. For example, one error out of 10^8 is expressed as a BER of 10^{-8}. A realistic estimate is placed on the bit error rate that can be tolerated in the transmission path, while retaining overall system specifications. For example, a transmission path that carriers voice or video will have different bit error rate requirements than a path that carries computer data. More stringent requirements such as BER = 10^{-9} require the use of error-correcting codes. In telecommunications, higher-level system protocols may be used to monitor errors and request retransmission when errors are detected. These factors must be considered when the optical link's BER performance requirement is established.

The actual design of a system entails evaluation of such variables as operating wavelength, type of fiber, emitter and detector, and the selection of specific transmitters, receivers, cables, connectors, and other components. Generally speaking, the best approach involves making initial assumptions and choices, and then performing an analysis (evaluation) of the tradeoffs.

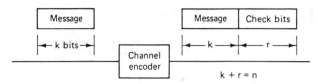

Figure 3.8 Example of a block encoder.

Data	4B/5B Code
0000	11110
0001	01001
0010	10100
0011	10101
0100	01010
0101	01011
0110	01110
0111	01111
1000	10010
1001	10011
1010	10110
1011	10111
1100	11010
1101	11011
1110	11100
1111	11101

Figure 3.9 4B/5B code.

3.2.1 Distance and Data Rate

The distance limit between transmitter and receiver (or between repeaters) is determined by the source power, the fiber loss, and the repeater sensitivity (see Section 3.1). Receiver sensitivity is a function of both the data rate and the required BER. We can divide the limits on system speed into two types: those that depend on the length of fiber and those that are constant. Examples of constant effects are the rise times of the transmitter and receiver. Fiber-length-dependent effects arise from pulse spreading in the fiber. Thus, the fiber is often characterized by its bandwidth-distance product. This permits a tradeoff between distance and data rate.

Many local fiber optic applications will operate in the 800- to 900-nm optical wavelength window, where good performance and relatively inexpensive components are available. Multimode fiber is the choice for local applications for the following reasons:

- It works well with relatively simple and inexpensive LED emitters and PIN diode detectors.
- It provides good attenuation and bandwidth performance in the 850-nm wavelength range.
- It provides even better performance in the 1300-nm range.

In approaching a design, the first parameter to be selected is the wavelength. The selection of a wavelength reduces the number of active components to be selected by about 50 percent. For distances less than 1 km, the preferred wavelength is 800 nm. Plastic fiber can be used for distances up to 50 meters, and step-index or graded-index fiber can be used for distances up to 1 or 2 km. An LED source can be used for most applications. Laser sources are available if higher data rates are required. Figure 3.10 illustrates the tradeoff between LED and laser components at both 800 and 1300 nm. Note that the longer wavelength provides both a higher maximum data rate and longer distance at low data rates.

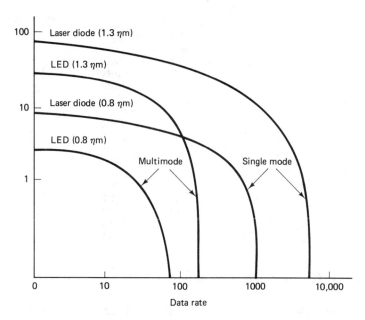

Figure 3.10 Lasers provide better performance than LED sources. Distance and data rate requirements are improved because of the laser's higher output power and smaller spectral width.

For links between 1 and 5 km, a laser in the 800-nm window or an LED in the 1300-nm window can be used. We can use graded-index fiber with the LED and single-mode fiber with the laser. For data rates greater than 1 Gbit/s, a wavelength of 1300 or 1550 nm is usually needed along with a single-mode fiber. A laser can be used with multimode fiber. In an existing installation, this allows upgrading the data rate by replacing the original LED with a laser.

For distances greater than 10 km, a laser source and a single-mode fiber are required. For reasons of economy, a wavelength of 1300 nm is preferred. Distances that are greater than 25 km require the wavelength choice to be 1550 nm. A dispersion-shifted fiber is required to achieve the best data rates.

In order to keep the cost down, overdesign of a system should be avoided. As the wavelength increases, so does the cost. A low-speed link, then, should not be designed with 1550-nm components. In an attenuated limited system, the designer has several choices:

- Choose a more powerful source.
- Choose a more sensitive detector.
- Use a lower-loss fiber.

If the system is bandwidth limited, the designer can select a fiber with less pulse spreading, select a longer wavelength, or replace the LED with a laser. The cost of each choice must be evaluated.

3.2.2 Selecting Components

Once the wavelength has been defined, the source and receiver are selected. From Chapter 2, we know that for linear systems, an LED and a PIN are chosen for their linearity. For digital applications, the choice is based on speed. System specifications identify appropriate transmitters and receivers based on signal characteristics such as data rate, duty cycle, and so on. An application may require 100-Mbit/s transmission with a given power budget and with a BER of 10^{-9}. A receiver specified at 100 Mbits/s and a BER of 10^{-12} may operate properly due to a tradeoff between data rate and bit error rate. Engineering analysis may be needed to resolve the issues.

Generally speaking, for speeds up to 10 Mbits/s, an LED transmitter is usually used. For speeds above 1 Gbits/s, a laser is preferred. Between these speeds, an LED is preferred for shorter links and a laser for longer links. The choice is based on fiber pulse spreading. The fiber is chosen to keep both the pulse spreading and attenuation within allowable limits. With an LED, the coupling efficiency to the fiber is equal to the square of the numerical aperture (see Example 1.4). For example, for an NA of 0.3, only 9 percent of the source power is actually coupled into the fiber. This assumes that an index matching gel is used [see Equation (1.10)]. Without index matching, the coupled power drops to around 1 percent. For a laser, the coupling is dependent on alignment between the connector and fiber cores. Typical losses are in the range of 50 percent.

Once the components are chosen, system design requires the analysis of signal integrity throughout the system. The interaction between optical power and the bandwidth must be continuously examined.

3.2.3 Power Budget

The analysis of a design proceeds in two phases. The designer must first check the power budget to ensure that power at the receiver is high enough that the receiver can determine if a "one" or "zero" is present. Next, the system rise time is analyzed to determine if the bandwidth required by the data rate is available. If results from this analysis are positive, the design is ready for installation.

The key elements in the analysis of a fiber optic link are

- Transmitter power out, P_T, into a specific fiber
- Fiber attenuation per unit link, A_L
- Receiver sensitivity (minimum input power), P_R
- Connector and component loss, A_C
- Link margin, M

We can use Equations (1.18) and (1.19) to express the link margin M as

$$M = P_T - P_R - A_L \cdot L - A_C \qquad (3.14)$$

where L is the fiber length.

Normally, when Equation (3.14) is used, peak power values are used. Sometimes average power, based on a 50 percent duty cycle, is specified for a transmitter or receiver. Either peak or average power values can be used, but they cannot be intermixed within a calculation, because an inconsistency results that causes sizable errors. As noted previously, connector loss, fiber attenuation, and length values, along with specifications for the transmitter and receiver, determine an acceptable link margin. Links that are expected to be extended in distance or to receive additional connectors and components will require more link margin for the increase in attenuation. An example will illustrate link margin calculations.

Example 3.2

An 86-micron core fiber with 5 dB/km attenuation is selected to transmit 100 Mbit/s data. From specifications, the launched power for the selected fiber is -16 dBm and worst-case receiver sensitivity is -30 dBm. Connector loss is 1.5 dB per mated pair, and two connector paths are anticipated in the link. Allow 5 dB component aging and/or link extension. (a) Compute the allowable link length, and (b) plot the power levels versus link length.

Solution

(a)
$$M = P_T - A_L \cdot L - A_C - P_R$$
$$5 = -16 - A_L \cdot L - 2(1.5) - (-30)$$
$$A_L \cdot L = 6$$
$$L = \frac{6}{5} = 1.2 \text{ km}$$

(b) Power levels are plotted in Figure 3.11 showing the connectors placed at arbitrary points in the link. This yields a graphical display of results in the power budget.

A degree of care is in order in interpreting and using the power budget. The attenuation for multimode fibers and cables is specified under particular optical launching conditions. A typical parameter test, referred to as the *cut-back method*, involves launching power into a 1- to 30-km length of fiber and then measuring the

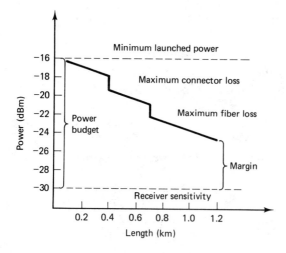

Figure 3.11 Power level vs. line length.

optical power at the other end. Next, most of the fiber is cut off and the optical power is measured at the output of the remaining short stub. The attenuation value (dB/km) attained is accurate for the particular length and launching optical conditions, but does not necessarily reflect the loss under other conditions. For example, when launched power fully covers the area and numerical aperture of the fiber core, a relatively large amount of power is contained in the higher-order modes. These modes are more susceptible to loss than the lower-order modes and can result in losses ranging from 1 to 1.5 dB when they are removed from the fiber. Thus, total power in the fiber drops quickly at first, and then more slowly, as the higher-order modes are dissipated. This higher-order mode loss is called *transient loss*. Loss due to the lower-order modes is called *steady-state loss*. Under conditions where a source fully excites the fiber modes, larger attenuation is experienced for short lengths than is predicted by the linear attenuation assumption of constant dB/km. These variations of optical power with link length are illustrated in Figure 3.12.

Connectors that are near the launching end of the link will tend to selectively attenuate the higher-order modes. These modes would have been attenuated in the fiber. On the other hand, connectors placed where a steady-state mode distribution exists may cause some power to be coupled into the higher-order modes, thus causing additional transient loss in the fiber.

In addition to the variations in fiber attenuation, the minimum receiver sensitivity is based on several variables, including actual data rate, required BER, and duty cycle. The maximum data rate specification for the receiver is dependent on the bandwidth of the preamplifier circuitry. At the maximum data rate, high-frequency roll-off of the amplifier attenuates the signal to some degree. At slower signal rates, the gain of the amplifier is greater, and less signal power is required to achieve the same results (see Figure 3.13). Receiver sensitivity at a particular BER is based on standardized or widely accepted values. However, particular applications may involve either higher or lower bit error rates. Figure 3.13 illustrates how signal power and BER vary for a typical receiver.

Duty cycle variations affect different types of receivers in different ways (see

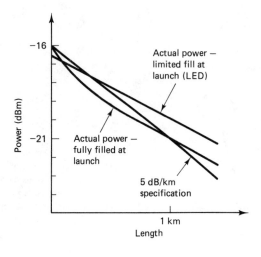

Figure 3.12 Loss curves illustrating constant attenuation in decibels per kilometer along with more accurate curves, which include both transient and steady-state loss.

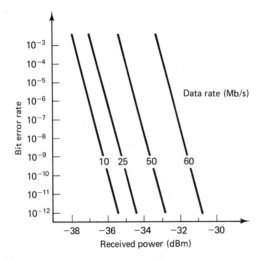

Figure 3.13 Sensitivity varies with data rate for a 50-Mb/s receiver. This information is used in applications having a data rate below the maximum specified for the receiver. At a particular level of BER, additional sensitivity will result. Applications having a data rate somewhat above the one specified for the receiver may still be considered, because the receiver may be used with an appropriate reduction in the sensitivity level.

Figure 3.14). This is particularly true for AC-coupled receivers. A DC-coupled receiver will handle any duty cycle without degrading the BER.

3.2.3.1 Required Optical Power. The required optical power varies with the BER and type of coding scheme chosen. The required optical power for a BER = 10^{-9} is plotted in Figure 3.15 versus the bit rate for both PIN and APD diodes. This assumes unipolar PCM and a bipolar preamplifier. The minimum average optical power is given by

$$P_{\min} = \frac{hf}{\eta e} Q \langle i_{\text{eff}}^2 \rangle^{1/2} \qquad (3.15)$$

where i_{eff} is the effective value of the noise current, consisting of shot noise, thermal noise, and transistor noise [see Equation (2.10)]. The quantity hf is equal to 2 ×

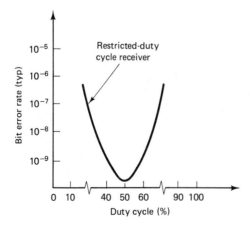

Figure 3.14 The degradation of BER can be plotted for a duty cycle that varies from 50 percent by using a receiver with a restricted duty cycle operating near its maximum rate. A DC-coupled receiver will handle any duty cycle without degrading the BER.

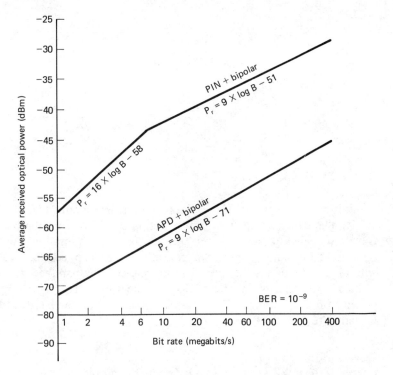

Figure 3.15 Required optical power versus bit rate.

$10^{-19}/(\lambda/\mu m)$. In this equation Q is a function of the error probability and may be expressed as

$$Q = \sqrt{2}\,\text{erfc}^{-1}(2P_e) \tag{3.16}$$

Equation (3.16) is plotted in Figure 3.16. For a given BER, we can obtain P_{min} if we know P_{min} and i_{eff}. Refer to Equation (2.2). The shot noise varies as the square root of the signal. Thus, the effective value of the noise current i_{eff} is relatively constant with changes in received optical power. An example will illustrate.

Example 3.3

Determine the minimum required received power needed to transmit data at a rate of 2 Mbits/s with a BER = 10^{-5}. $\lambda = 0.8$ μm and a PIN diode is used with 60 percent efficiency.

Solution From Figure 3.15, the required received power is

$$\begin{aligned} P_r &= 16 \log B - 58 \\ &= 16 \log 2 - 58 = -53.18 \text{ dBm}(4.8 \times 10^{-6} \text{ mW}) \end{aligned}$$

From Figure 3.16, $Q = 6$ for BER = 10^{-9}. The effective noise current is

$$\langle i_{eff}^2 \rangle^{1/2} = \frac{P_{min}}{Q(hf/\eta e)}$$

Sec. 3.2 Designing Digital Fiber Optic Links

$$= \frac{4.8 \times 10^{-9}}{6[(2 \times 10^{-19}/0.8)/(0.6)(1.6 \times 10^{-19})]}$$

$$= 0.3072 \times 10^{-9} = 0.3072 \text{ nA}$$

For $P_e = 10^{-5}$, $Q = 4.2$.

$$P_{min} = \frac{(2 \times 10^{-19}/0.8)(4.2)(0.3072 \times 10^{-9})}{(0.6)(1.6 \times 10^{-19})}$$

$$= 3.36 \times 10^{-6} \text{ mW}(-54.73 \text{ dBm})$$

3.2.4 Bandwidth Budget

In addition to power, the system rise time must be examined (see Chapter 2, Section 2.5.1). The data rate capability of a link is obtained from the bandwidth budget. Generally speaking, the receiver is the most bandwidth-limiting component in a fiber optic link. In many designs, the transmitter, interconnection hardware, and fiber will not greatly affect the overall performance. The bandwidth budget is a useful technique for determining these effects of the transmitter and the other system components on link data rate. The system rise time t_r and the bandwidth are related by Equation (2.61). That is,

$$T_{system} = t_r = 0.35\tau \tag{3.17}$$

where τ is the pulse width. For an RZ code, the bit rate $R = BW = 1/\tau$ (see Chapter 4). Using Equation (3.17), this results in

$$R_{max} = \frac{0.35}{T_{system}} \tag{3.18}$$

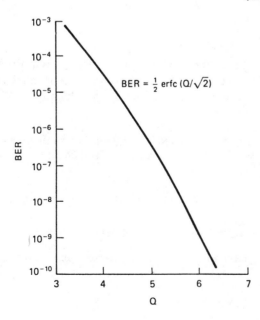

Figure 3.16 BER versus Q for binary PCM reception.

On the other hand, for an NRZ code, $R = BW/2 = 1/2\tau$ and R_{max} is given by

$$R_{max} = \frac{0.7}{T_{system}} \qquad (3.19)$$

Example 3.4

For the fiber optic system of Example 2.7, estimate the maximum bit rate that the link will support for (a) RZ and (b) NRZ.

Solution

(a) $R_{max} = \dfrac{0.35}{T_{sys}} = \dfrac{0.35}{46.17 \times 10^{-9}} = 7.58$ MHz

(b) $R_{max} = \dfrac{0.7}{46.17 \times 10^{-9}} = 15.16$ MHz

We see from Example 3.4 that the information-carrying capability of the fiber optic link can be determined in the time domain. Pulse broadening analysis is oriented toward rise time and fall time. The term *transition time* refers collectively to the rise time and fall time. It should not be confused with propagation (transit) time—the time it takes signals to pass through a fiber. Equations (3.18) and (3.19) are empirical rules that are used to check system bandwidth. The effective transition time t_s of the optical signal entering the receiver should be less than 70 percent of the signal bit time. That is,

$$t_s = 0.7T \qquad (3.20)$$

where T is the signal bit time. For NRZ

$$t_s = 0.7/B \qquad (3.21)$$

where B is the bandwidth in hertz of the NRZ data.

As long as t_s is shorter than this value, the full system bandwidth is limited by the receiver. If t_s is longer, the performance of the link is limited by an element other than the receiver. When the signal has a long transition time, the BER level will increase. That is, the sensitivity of the receiver will decrease for a given BER value (see Figure 3.13). The major items that combine to lengthen (degrade) transition time are (1) transition time of the transmitter including the emitter t_e, (2) transition time due to modal dispersion t_m, and (3) transition time due to chromatic dispersion t_c. In general, connectors and splices are not included because they do not limit the system bandwidth. In fact, these devices may improve system performance somewhat by causing mode mixing.

We use the square root of the sum of the square formula to obtain a value for t_s. That is,

$$t_s = (t_e^2 + t_m^2 + t_c^2)^{1/2} \qquad (3.22)$$

As noted previously t_e is due to the transmitter electronics and transmitting diode (laser or LED). The model dispersion can be calculated from

$$t_m = D_{mod} \cdot L^Q \qquad (3.23)$$

where

L = length of the fiber, km

Q = a constant that indicates how the modal dispersion scales with length

D_{mod} = amount of modal dispersion for a particular fiber, ns/km

A reasonable value for Q in very long fibers is 0.7. For short fibers, an estimate for Q is 1.0. D_{mod} indicates the amount of modal dispersion of a particular fiber. Since this is nonlinear with length, it is important that a value for D_{mod} be obtained from a known fiber length.

If a value for D_{mod} is not available, the modal dispersion should be estimated, based on the bandwidth specification B_F for the fiber. A technique for determining the bandwidth is illustrated in Figure 1.8. Since the transition time, and thus bandwidth, may not scale linearly with the length, specifying bandwidth B_F in units of MHz•km is not a precise indication of actual performance. These bandwidth values are valid only if obtained on a 1-km length. If the bandwidth is obtained with a different fiber length L_0, the transition time due to modal dispersion is given by:

$$t_m = 0.44L^Q/B_F$$
$$= 0.44/B_F \cdot (L/L_0)^Q \qquad (3.24)$$

Example 3.5

(a) A fiber has a bandwidth of 200 MHz•km at 1 km. Calculate the transition time for a 300 meter length of the fiber.

(b) What is the transition time for a 2-km length?

Solution

(a) For short lengths, $Q = 1.0$.

$$t_m = 0.44L^Q/B_F$$
$$= (0.44)(0.3)(200 \times 10^6) = 0.66 \text{ ns}$$

(b) For long lengths, $Q = 0.7$.

$$t_m = 0.44(2^{0.2}/200 \times 10^{-6}) = 3.57 \text{ ns}$$

Waveguide dispersion may be ignored in multimode fibers in the 800-to-900-nm wavelength range. Material dispersion does exist, however, and manifests itself as chromatic dispersion in the fiber (see Section 1.1.2.2). In this wavelength range, common glass fibers have material dispersion D_{mat}, as illustrated in Figure 3.17. For positive values of D_{mat}, the chromatic dispersion may be expressed as

$$t_c = D_{mat} \cdot \Delta\lambda \cdot L \qquad (3.25)$$

where

D_{mat} = the spectral width of the source, ns/nm•km

$\Delta\lambda$ = spectral width of the source, nm

L = fiber length, m

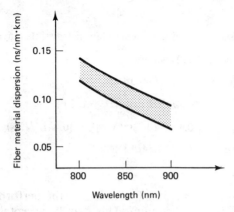

Figure 3.17 Values for D_{mat} in the literature are sometimes shown as positive and sometimes negative. A positive value is required for calculations.

Example 3.6

Calculate the chromatic dispersion for a system using an LED with a 50-nm spectral width which drives a 300-m length of fiber. The fiber has a material dispersion of 0.1 ns/nm/km.

Solution

$$t_c = D_{mat} \cdot \Delta\lambda \cdot L$$
$$= (0.1 \text{ ns/nm/km})(50 \text{ nm})(0.3 \text{ km}) = 1.5 \text{ ns}$$

The spectral width of LEDs is relatively large compared to laser diodes, and material dispersion in LED systems operating in the range of 850 nm often will be more limiting than modal dispersion.

From the discussion in Chapter 1, we know that material dispersion at 1300 nm is very low. At this wavelength, waveguide dispersion should be considered, and an accurate value for modal dispersion is important. In this instance, the spectral width of an LED source leads to more complicated calculations than are illustrated by Example 3.6. At this point, using the information available, we can estimate the optical transition time t_s into the receiver. Assuming that the transmitter transition time t_e is on the order of 4 ns and using results from Examples (3.5) and (3.6), we have

$$t_s = (4^2 + 0.67^2 + 1.5^2)^{1/2} = 4.3 \text{ ns}$$

Now for a 50-Mbit/s receiver, the permitted transition time into the receiver is given by

$$t_s = 0.7/B = 0.7/50 \times 10^6 = 14 \text{ ns}$$

Thus, for the system values chosen, the receiver performance is not degraded at 40 Mbits/s. We can readily determine the maximum length of the fiber that can be used before t_s exceeds the specified limit, and begins to degrade system performance.

Example 3.7

For the system discussed, determine the maximum permissible length of fiber.

Solution Using Equation (3.22), we obtain

$$t_s = (4^2 + t_m^2 + t_c^2)^{1/2} = 14 \text{ ns}$$

$$t_s^2 = [16 + (0.44L^Q/B_F)^2 + (D_{mat} \cdot \Delta\lambda L)^2]$$

$$(14 \times 10^{-9})^2 = [16 + [(0.1 \times 10^{-9})(50)(2)]^2 + [0.44L^{0.7}/200 \times 10^{-6}]^2$$

$$196 = 16 + 25L^2 + 4.84L^{1.4}$$

$$L = 2.5 \text{ km}$$

If t_s exceeds the limits indicated in Example 3.7, the performance of the receiver will be somewhat less than optimum. This may be acceptable if the link is operating at a data rate lower than the maximum data rate specified for the receiver. The system can benefit from increased sensitivity at the lower data rate. Example 3.7 indicates system tradeoffs that can be effected if improvements in optical transition time is needed. For starters, we may consider a faster transmitter. Material dispersion is reduced by moving to the 1300-nm wavelength or by using an LED with reduced spectral width. Note that in this example, modal dispersion has the greatest effect on transition-time degradation. A fiber with a larger bandwidth may offer some improvement.

The calculations just considered involve a large number of approximations and empirical estimations. If this results in a borderline system design, more exact measurements are dictated. For example, performance measurements, including power and BER studies, are needed for the actual system capability.

3.2.5 Optical Transmitter and Receiver Considerations

Link components contain sensitive high-speed transistors and other circuitry. Thus the devices must be protected from electrostatic discharge through proper handling and grounding techniques.

For an initial evaluation of transmitters and receivers, the designer can use evaluation kits. Such kits include printed circuit boards that provide a convenient solution for putting the devices into operation for engineering evaluation. Evaluation board schematics, included in data sheets or drawings in the kit, provide guidelines for implementing custom boards. The transmitters and receivers are similarly packaged. However, they have substantially different requirements.

The transmitter and receiver modules should be located near the edge of the circuit board to avoid routing the fiber over the board and to allow easy access for fiber connection. This location helps prevent other circuitry from injecting noise into the receiver module. Link modules should be located on double-layer or multi-layer circuit boards only. A ground plane should be placed under the full area of the module. No circuit traces should be routed between the ground plane and the receiver module; however, if necessary, signals may be carefully run near the module on the side of the board opposite the ground plane. Care should always be exercised in multiboard systems to prevent strongly interfering signals on another board from being physically close to the receiver module.

Data line connections are highly dependent on the speed of the signals. Modules with ECL data connections need circuit traces that conform to nominal ECL wiring practice. This will generally require the use of microstrip or stripline wiring on the PC board (Figure 3.18). Optical transmitters require substantially more power than typical integrated circuits, and suitable conductor traces should be provided. Transmitters normally do not require special cooling, and placement of the module close to sources of heat, which may require care in maintaining the module package temperature at acceptable levels; should be avoided.

Circuit boards that are supplied in evaluation kits pay particular attention to power supply decoupling and filtering requirements. Custom applications should equally emphasize the importance of this area. Where two connections are indicated for a particular voltage, such as $V_E E1$ and $V_E E2$, separate power supply connections are warranted. The $V_E E1$ connection to the preamplifier of the device is sensitive to the conducted interference. A pi-section filter is recommended on the supply pin, particularly on the higher-speed link products. High-grade capacitors suitable for RF use should be employed.

A low-resistance, low-inductance ground path for the modules must be provided. It is important that all designated ground pins be well grounded to maintain proper operation.

Data connections should be microstrip or stripline geometry. The characteristic impedance of the lines can be chosen to be consistent with other circuit elements (see Figure 3.18). Where the lines are of significant length, or the modules are connected to cables off the board, transmission line terminations will maintain signal integrity. ECL transmitters are normally driven differentially. If a single-ended drive is desired, the unused input must be connected to the V_{BB} pin of the transmitter and capacitively bypassed to ground.

Link modules can be interfaced to a wide variety of common integrated circuits. It is recommended that a single gate be dedicated as the interface between

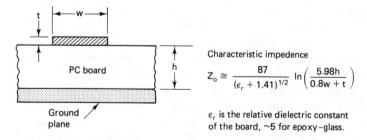

Characteristic impedence

$$Z_o \cong \frac{87}{(\epsilon_r + 1.41)^{1/2}} \ln\left(\frac{5.98h}{0.8w + t}\right)$$

ϵ_r is the relative dielectric constant of the board, ~5 for epoxy-glass.

Figure 3.18 A typical microchip geometry cross section may serve as a reference for board design. Modules with TTL data connections will require typical TTL practices. Techniques similar to those used with ECL may be required with the highest-speed TTL families of logic, since the signal rise times are almost equal to those of ECL.

each module and the rest of the digital circuitry. The gate will act as a buffer to help isolate the modules from other circuit elements and prevent unwanted interaction.

In addition, transmitter and receiver modules should not require adjustment or alignment in operation. The optical interfaces should be reliable and trouble free through their service life. Normal care should be exercised so that the optical connectors and optical ports of the modules are kept clean and free from debris. When needed, cleaning should be done with optical-grade, lint-free materials. Versatile transmitters will permit straightforward application of the module in most applications. DC-coupled transmitters are insensitive to data format or frequency content, up to the maximum bit range specified. No particular precautions or limitations should be imposed on the signals presented to the transmitter, except for logic level and transition time constraints.

AC-coupled receivers, rated at 50-, 100-, and 200-Mbits/s, provide restricted-duty-cycle operation. Optimum performance of the receiver occurs with signals having an average duty cycle near 50 percent. This type of receiver takes advantage of the statistics found in commonly encoded data streams, such as those produced with Manchester encoding or data scrambling or encryption circuits. Such receivers employ decision circuitry having a threshold at the DC (average) level of the incoming signal. When signals with very small or very large average duty cycles are presented to a restricted-duty-cycle receiver, the decision threshold shifts toward one of the signal levels, increasing the probability that noise will contaminate the signal. This results in increased bit-error or reduced sensitivity for a given BER (see Figure 3.14).

Restrictions placed on the duty cycle of the data entering the receiver are based on certain assumptions about the overall performance of the module. Operation at more extreme duty-cycle levels is possible, but performance tradeoffs will be needed. This may require ensuring power levels above the normal sensitivity limits of the receiver, or allowing higher levels of BER.

One particular limitation of restricted-duty-cycle receivers is their inability to perform well in situations where data occurs sporadically, in bursts. This implies long periods without signal transition between bursts, as well as deviation from a 50 percent average duty cycle. When data occurs on a link after such a period, the receiver will require sufficient time to stabilize at the average DC level produced by the data. Specifications will be met only after this period of stabilization.

DC-coupled receivers operate with arbitrary data streams, with no restrictions on the data duty cycle, encoding, or transmission bursts. The device is sensitive to the transition of the signal, whereas the restricted-duty-cycle receiver uses level sensing. These devices are simple in design and relatively easy to apply. The edge sensing that is typically used in a DC-coupled receiver is not a particularly high-performance approach to receiver design. This architecture often results in 6-to-8-dB penalty in the sensitivity of the device, compared with the restricted-duty-cycle design. Consequently, it is more suited to optical links with limited distance, low data rate, or modest error-rate specifications.

Handling data without regard to duty cycle or format can be implemented with the high-performance, restricted-duty-cycle receiver designs. These employ encoding circuits in the transmitter module, decoding, and clock recovery circuits in the receiver, to automatically transmit and receive data without regard to coding or duty cycle. These circuits perform their function while being transparent to the user. The user connects data and clock signals to the transmitter, and obtains data and clock signals from the receiver.

This approach finds many applications in high-performance links, where DC-coupled devices are not adequate. Many applications require clock recovery at the receiver end of the link, so this approach satisfies this needed feature. Those products that are currently available are usually in the form of costly board-level modules, rather than components. However, the trend toward this type of component is being driven by the performance advantages that it offers.

3.3 TESTING THE DESIGN

Testing of the completed design can be accomplished in a number of ways. Among these are

- Power at the source and at the fiber end (output) can be measured.
- Time-domain measurements such as an "eye pattern" can be made with an oscilloscope.
- Spatial-domain measurements on the fiber can be made with an optical time domain reflectometer (OTDR).
- Data-domain measurements are made with a BER tester.

Power meters are used to measure the DC optical power. The key points of interest are the source output power and the fiber output power. Measurement of the source power should be made at both component-qualification time and during troubleshooting of the link. Power measurement at the fiber end can be used to determine fiber attenuation, link power loss, connector losses, and ultimately, power available to the receiver.

For a purely digital system, a bit error rate (BER) tester is used to measure the BER to determine data quality. In an installed system, the measured BER must be less than the required BER. In some instances, monitoring the BER during operation will allow a decrease in the source power to be detected before the link fails completely.

An optical time domain reflector (OTDR) can be used to characterize fiber links in place (see Chapter 8). These devices operate by transmitting a signal and monitoring the returns. OTDRs are used to identify fiber breaks and to check loss along the installed link. The OTDR allows measurements to be made from the sending end of the fiber. Conversely, BER testers and power meters normally require simultaneous access to both ends of the fiber.

The receiver output for systems that are bandwidth- or rise-time-limited can be checked with an oscilloscope by observing the "eye pattern" and system rise time. An eye pattern is generated by driving the optical transmitter with a pseudo-random data generator. The output of the optical receiver is connected to the vertical input of the oscilloscope, and the oscilloscope is externally triggered from the source (data generator). The horizontal sweep time of the scope is set to approximately one bit period. The eye pattern, then, is the superposition of many data transitions (see Figure 3.19). Some key eye-pattern analysis techniques are

- Sample the data at the maximum vertical opening of the eye.
- The period the data can be sampled is shown as the width of the central eye opening.
- The noise margin of the receiver output is the vertical height of the central eye opening (optimum sampling point).
- The slope of the signal lines on either side of the opening indicates the timing sensitivity of the data sampling point. A fully open center is preferable to one that closes rapidly with small changes from the optimum sampling time.
- The width of the signal band at the horizontal line midway between minimum and maximum signal levels represents timing jitter. Jitter represents variations in the signal switching point that can adversely affect system timing (changes zero-crossings).
- Noise and distortion are indicated by the vertical thickness of the signal band at the optimum sampling point. Increased thickness is undesired.
- Signal rise and fall times can be measured from the eye pattern by observing the signal transitions between the bottom and top of the pattern, and then determining the 10–90 percent or 20–80 percent transition times.
- The central rectangle can be established for a particular application to indicate signals that are acceptable in amplitude and timing. Signals are acceptable if the eye pattern falls outside the window that has been established.
- A large eye pattern signals an acceptable signal-to-noise ratio.

Checkout and testing of fiber optic systems are considered in more detail in Chapter 8.

3.4 DATA BUS TOPOLOGY

We briefly mentioned the architecture of fiber optic systems in Chapter 1 (see Section 1.3.2). We noted that T (ring) and star networks are most common. A T net-

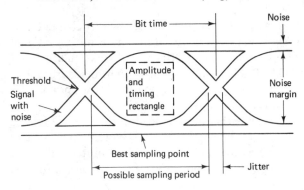

Figure 3.19 Typical eye pattern results from the superposition of many data transitions. Qualitatively, a wide-open eye with symmetrical appearance is desired, so as to avoid intersymbol interference (ISI) and data errors.

work for N stations is illustrated in Figure 3.20. If we visualize station 1 as transmitting into the fiber, we see that the remaining networks remove power from the network. The power available at station $N - 1$ can be expressed as

$$\frac{P_{N-1}}{P_1} = L_{LS} + (2L_C + L_{INS} + L_{IT})(N - 2) + 4L_C + L_{INS} + L_{TR}, \quad N > 3 \quad (3.26)$$

where

L_{LS} = fiber coupling loss at the light source

L_C = connector loss

L_{INS} = insertion loss of each coupler

L_{IT} = $-10 \log_{10} [1 - 10^{L_{TR}/10}]$ is the insertion loss of each coupler due to a tap

L_{TR} = tap ratio loss at the $(N - 1)$ coupler

Fiber optic T couplers may be used to form a network of stations, as illustrated in Figure 3.20. They are limited to applications involving a limited number of stations.

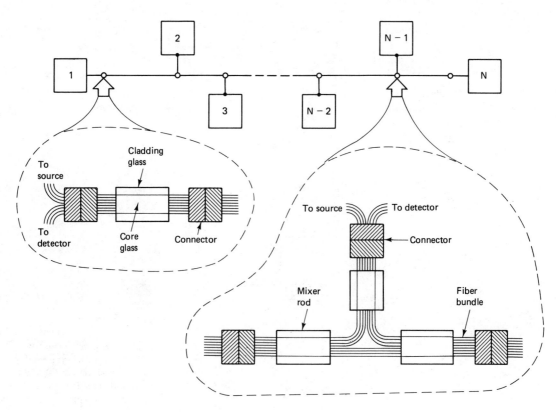

Figure 3.20 An N-station T system.

Sec. 3.4   Data Bus Topology

Although T couplers are available with different splitting ratios, they all remove a significant excess amount of power from the network. Since a T coupler may be required for each node in a network, the losses increase greatly as more couplers are added. In addition, dynamic range problems occur. Nodes near the source are subject to larger amounts of power than are nodes that are farther away.

Some networks may use a star coupler to advantage in creating a broadcast-style network. A star coupler is illustrated in Figure 1.28 and is repeated in Figure 3.21. The relationship between the input and output powers through any two ports on the star coupler is given by

$$\text{Coupler loss} = 2C + E + 10 \log N \tag{3.27}$$

where

C = optical connector loss (typically 1.5 dB)

E = star coupler "excess loss" (typically 1 to 4 dB)

N = numbering optical parts

Note that the splitting loss of star couplers increases logarithmically as the number of coupler ports increases. Excess loss, the additional loss that results because a

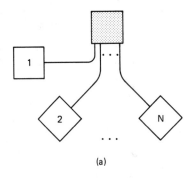

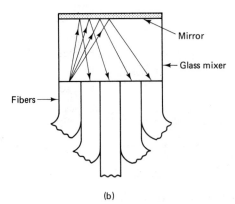

Figure 3.21 (a) One terminal can communicate with all others when you employ the star-coupler configuration. (b) Light entering at one port spreads out in the glass mixer and reflects from the mirror into all other ports.

coupler is not an ideal device, occurs only once with a star-coupled network, whereas it occurs at every point in a T network. A star system is illustrated in Figure 3.22, and a multiterminal communication system employing a star coupler is illustrated in Figure 3.23.

The power into any port of a star coupler is nearly equally divided among all output ports. Because loss increases slowly with the number of additional output ports, the penalty paid for using a coupler larger than required, with ports reserved for future expansion, is balanced against the advantages of easy expansion by simply adding stations. The power budget for existing stations is not affected by new stations. In a network involving mainly transceivers, the star coupler is the favored solution. Combinations of star and T couplers may prove to be the optimum solution to network flexibility.

In applications where a signal must be split or combined from one fiber into many or at an emitter or detector, a splitter/combiner can be used. A splitter is shown in Figure 3.24. Note that it bundles several fibers together at one end. We can express the ratio of power at any port to the source power as:

$$\frac{P_{N-1}}{P_{LS}} = L_{LS} + 4L_C + L_{INS} + L_{TR} \qquad (3.28)$$

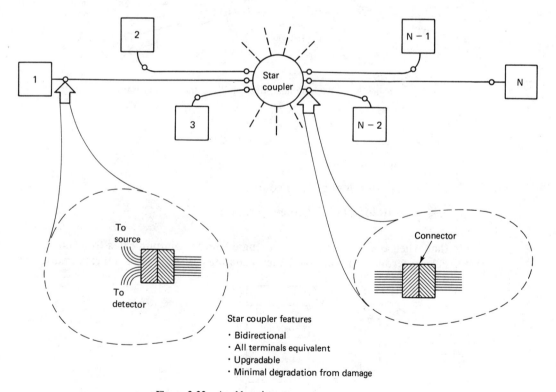

Figure 3.22 An N-station star system.

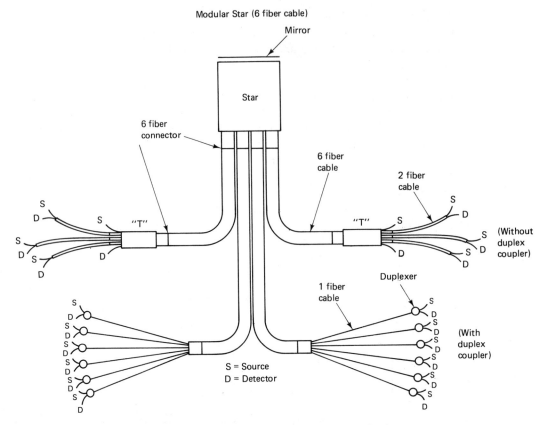

Figure 3.23 Multiterminal fiber optic communication system.

where

L_{LS} = fiber coupling loss at the light source

L_C = connector loss

L_{INS} = insertion loss of each coupler

$L_{TR} = -10 \log_{10}(1/N)$ is the tap ratio loss

Note that when the splitter is used, a single source communicates to a number of receivers. In terms of a combiner, the availability of large-area sources and detec-

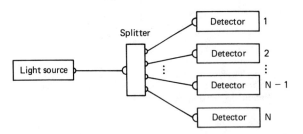

Figure 3.24 Fiber optic splitter.

108 Digital Fiber Optic System Design Chap. 3

tors, as well as of large-core fibers, permits each fiber in a bundle to communicate with a single fiber, emitter, or detector.

The number of terminals in a network strongly influences the choice between star and T couplers. A comparison of the two devices is illustrated in Figure 3.25. Considerable flexibility is available in network architectures. Additional architecture schemes are illustrated in Figure 3.26.

In designing fiber optic data communication networks, two criteria must be satisfied in order to ensure reliable optical performance. These are

- Sensitivity limit
- Dynamic range limit

The optical losses (attenuation) due to the coupler, optical fiber attenuation, any patch panel connectors or splices, together with the operating margin, must not cause the signal launched into the fiber (at the transmitter) to be attenuated below the level at which it can be detected at the receiver. This is generally set by the desired BER. At the same time, optical receivers have an upper limit to their detection range, above which they saturate. The optical circuit must contain sufficient

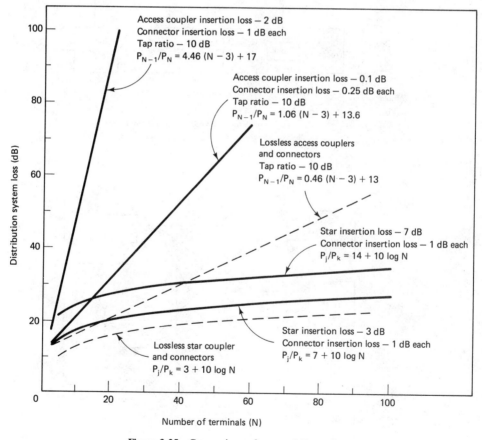

Figure 3.25 Comparison of star and T couplers.

Sec. 3.4   Data Bus Topology   109

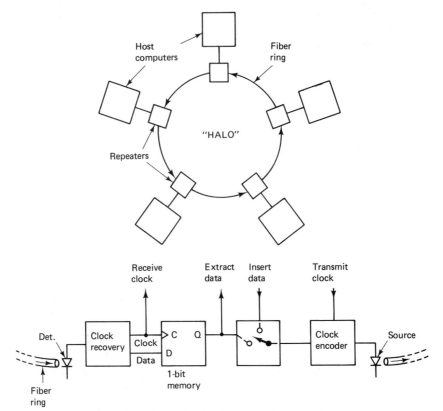

"Halo," a unidirectional repeatered loop network.

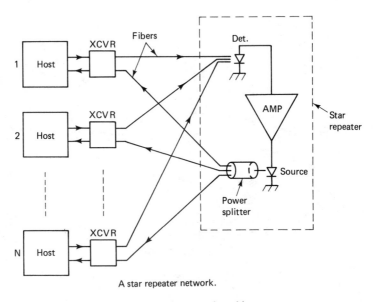

A star repeater network.

Figure 3.26 Fiber optic network architectures.

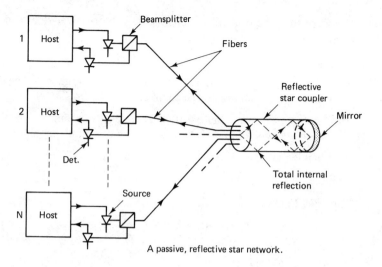

A passive, reflective star network.

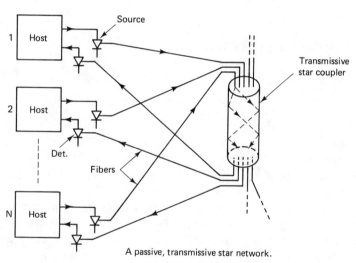

A passive, transmissive star network.

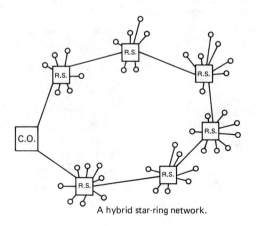

A hybrid star-ring network.

Figure 3.26 (*continued*)

loss to ensure that the optical power reaching the receiver does not exceed the dynamic range of the receiver.

Example 3.8

(a) A star coupler with 32 ports is available. Based on the preceding discussion, calculate the maximum loss of the coupler.

(b) Assuming a system gain of 30 dB and a 5-dB/km multimode fiber, calculate the maximum coupler-to-node separation. An operating margin of 4 dB is required.

(c) What is the required receiver dynamic range?

Solution

(a) Using Equation (3.27), we have

$$\text{Coupler loss} = 2C + E + 10 \log N$$
$$= 2(1.5) + 4 + 10 \log 32 = 22.05 \text{ dB}$$

(b) $M = \text{gain} - \text{losses}$
$4 = 30 - 22.05 - (5 \text{ dB/km})L$
$L = 0.79 \text{ km}$

(c) Dynamic range $= \text{gain} - \text{coupler loss}$
$= 30 - 22.05 = 7.95 \text{ dB}$

REFERENCES

3.1. BORLEY, D. R., et al. 1984. Some design aspects of long-haul digital submarine optical fibre systems. *The Radio and Electronic Engineer* 54: 163–172.

3.2. LEWIS, J. R. 1984. Factors involved in determining the performance of digital satellite links. *The Radio and Electronic Engineer* 54: 192–198.

3.3. LILLY, C. J., and WALKER, S. D. 1984. The design and performance of digital optical fibre systems. *The Radio and Electronic Engineer* 54: 179–191.

3.4. HENRY, P. S. 1988. Introduction to lightwave transmission. *IEEE Communications Magazine* 23: 12–16.

3.5. HATFIELD, W. B., et al. 1988. Fiber optic LANs for the manufacturing environment. *IEEE Network* 2: 70–74.

3.6. SOUTHARD, R. K. 1987. Fiberoptic system design. *Fiberoptic Product News* 2: 12–19.

PROBLEMS

3.1. An available fiber has a propagation product specification of 1 Gbit/s·km. Assume that the communication system operates at its maximum possible transmission rate. **(a)** How long does it take to transmit 100 Mbits of data over the following distances: 5 km; 25 km; 200 km; 1000 km? **(b)** What wavelength should be used for each of the distances in **(a)**?

3.2. What is the maximum possible transmission distance without a repeater for the following optical system? Solve for both a PIN diode and an APD diode.

Bit rate = 2 Mbits/s

P_s = 1.5 mW

Combined coupling loss for the input and output of 4 dB

Average splice losses of 1.25 dB/km

Fiber attenuation loss of 3 dB/km

  (b) The fiber has indices of refraction of $n_1 = 1.48$ and $n_2 = 1.46$. Fiber diameter is 50 μm. Calculate the fiber bandwidth for the system, using both the APD and PIN diodes.

3.3. The rise time of a low-pass system is defined as the time required for the output to go from 10 percent to 90 percent of its final value when the input is a unit step. Find the rise time for **(a)** a single time constant RC low-pass filter with half-power bandwidth $W_{1/2}$ and **(b)** an ideal low-pass filter with bandwidth W.

3.4. A digital fiber optic link employs ideal binary signaling at a 30 Mbit/s rate and operates at a wavelength of 1.3 μm. A germanium photodiode detector is used with a quantum efficiency of 60 percent at this wavelength. If the BER at the receiver drops below 10^{-5}, an alarm is activated. Calculate the minimum optical power required at the receiver in order to keep the alarm inactivated.

3.5. An optical system is designed with an LED transmitter that launches an average optical power of 1 mW into a fiber at a wavelength of 0.8 μm. Cable attenuation is 2 dB/km. If the receiver requires 1000 photons in order to register a binary "one" with a 10^{-10} BER, calculate the following: **(a)** maximum transmission distance (without repeaters) when the transmission rate is 2 Mbits/s; **(b)** maximum transmission distance when the transmission rate is 2 Gbits/s.

3.6. For a binary polar NRZ signal, the bit error rate is given by

$$\text{BER} = \tfrac{1}{2}\text{erfc}(E_b/N_0)^{1/2}$$

where erfc stands for the complementary error function. Plot a graph of BER versus E_b/N_0. Use the table for erfc in the Appendix and $E_b/N_0 = 10$ dB.

3.7. A fiber optic system has the following parameters:
- Modal dispersion = 2 ns
- Chromatic dispersion = 4 ns
- Receiver transition time = 5 ns

Estimate the maximum data rate for **(a)** RZ and **(b)** NRZ.

3.8. (a) Prove Equation (3.26) by drawing and labeling a network for $N = 6$ with all appropriate losses.

  (b) Design the network so that the maximum dynamic range of any of the receivers is 7 dB. Use attenuators as necessary and determine the transmitter power needed at station 1. The following parameters apply:

$$L_{LS} = 1.0 \text{ dB}$$

$$L_C = 1.5 \text{ dB}$$

$$L_{INS} = 2 \text{ dB}$$

$$L_{TR} = 10 \text{ dB}$$

$$\text{Receiver sensivity} = -30 \text{ dBm}$$

4

Baseband Coding for Fiber Optics

4.0 INTRODUCTION

Digital systems, unfortunately, make errors. This tendency is generally described as the bit error rate (BER) of the system. For example, a BER = 10^{-8} is usually considered good and 10^{-3} is poor. The term *error rate* is defined as the number of errors in a given group of bits, divided by the number of bits, or the ratio of bits in error to transmitted bits in some time interval. The classical model of a digital system is shown in Figure 4.1. A transmitted pulse stream is corrupted by the addition of white Gaussian noise, and the regenerator (decision circuit) reconstructs the pulse stream as best it can from the noisy signal. The ability of this circuit to do so depends on the received signal-to-noise ratio. In this system the probability of a bit's being in error does not depend on whether one or more of the preceding bits was in error. Thus, the system can be reasonably well described by its "bit error probability" and it is correct to replace that term by the BER. Real-world systems, however, may be subject to such phenomena as "burst errors." In this case, an "averaging" interval should be specified with the BER.

Coding of data, prior to transmission, is undertaken for these reasons:

- To increase the information capacity of the system
- To enable error detection and possibly error correction to be performed
- To alleviate transmission problems

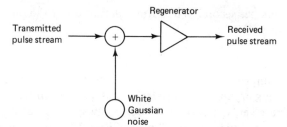

Figure 4.1 Classical digital system model.

Based on a purely theoretical approach, all three reasons should lead to the same coding scheme. This one ideal coding scheme, however, has not been realized (or even approached, for that matter) and consequently many different coding schemes exist. In 1948, C. E. Shannon developed a comprehensive theory that indicated the information capacity that could be obtained with optimum coding in a band-limited channel with noise. This paper, in fact, developed the fundamental thesis of information theory: All technical communications are essentially digital. More precisely, all technical communications are equivalent to the generation, transmission, and reception of random binary digits or "bits." That this is true for analog transmission is well documented. For example, we know that if a time-continuous (analog) signal has finite bandwidth, then it is completely specified by samples taken from it at discrete time instants, i.e., the sampling theorem. Thus, according to Shannon's theory, the discrete case forms a foundation for the continuous and mixed cases. This is a complete reversal of the roles of digital and analog communications in the earlier theory of communications.

One of the most surprising and deepest results of Shannon's theory was his demonstration that the problem of transmitting an information source through some channel can be separated, with no loss of optimality, into the independent problems of representing that source by a sequence of binary digits and of transmitting a random binary sequence through that channel. This is true even if both the source and channel are analog in nature. This is known as the *separation theorem,* and it has vast practical applications. Basically, the separation theorem means that communication transmission systems can and should be designed to transmit random binary digits (bits), and the resulting system can be used, without loss of optimality, to transmit any kind of information source such as analog. An example of the cost of not using this approach is illustrated by the efforts of communication engineers to transmit data over telephone (analog) channels today. A severe economic price is being paid for not adhering to the principles of the separation theorem. The advent of fiber optics represents a new opportunity for design.

4.1 SOURCE AND CHANNEL CODING THEOREM

Two of the greatest results of Shannon's theorem were the source (or "noiseless") coding theorem and the channel (or "noisy") coding theorem. The source coding theorem (noiseless) asserts that every communication source is completely characterized by a number H, the source rate. That is, the source is equivalent to one that emits H random binary digits per second. The formal definition of the quantity of information H is defined to be

$$H = \log N \tag{4.1}$$

where N is the number of equiprobable choices. This definition of H can be seen to be reasonable by the following argument: Consider a message transmitted using the English alphabet. At the receiver, the correct answer is always one out of the 26-letter alphabet or a space, for a choice of 1 of 27 possibilities. We can assign to each letter a number that represents its information content. A logical basis is the number of choices that the receiver must make, or the information value of each letter is 27. Next, consider a 2-letter combination. The number of choices is 27^2. For 3-letter combinations, the number of choices is 27^3. This approach leads to large numbers. It seems reasonable that a 3-letter combination carries three times the information of a single letter. This indicates that the information content varies according to a logarithmic scale. For a binary system, the number of choices N is two. If a value of 1 is desired as our basic unit of information, the base of the logarithm must also be 2. We can rewrite Equation (4.1) as

$$H = \log_2 N \text{ bits} \tag{4.2}$$

where the quantity of information H is expressed in bits and N is again the number of equiprobable choices.

The channel (noisy) coding theorem states that any communication channel is completely characterized by a single number, C, the channel capacity. More precisely, Shannon demonstrated that the system capacity C for channels perturbed by additive white Gaussian noise (AWGN—see Figure 4.1) is a function of the average received signal power S, the average noise power N, and the bandwidth W. This capacity relationship is known as the Shannon–Hartley theorem and may be expressed as

$$C = W \log_2 \left(\frac{S}{N} + 1 \right) \tag{4.3}$$

Shannon's theorem for noisy channels states that it is possible to transmit information over such a channel at a rate R, where $R < C$, with an arbitrarily small error rate by using a sufficiently complicated coding scheme. If $R > C$, it is impossible to find a code that can achieve an arbitrarily small error rate. Shannon's work demonstrated that values for S, N, and W set a limit on the rate, not on accuracy. This theorem demonstrated that one should not transmit information one bit at a time, but rather one should code long sequences of bits into a channel input sequence so that each bit of information is spread thinly over many channel uses. This idea gave birth to what is now the field of coding theory.

It is convenient to represent Equation (4.3) in graphical form. We can rewrite Equation (4.3) as

$$C = W \log_2 \left(1 + \frac{S}{N} \right)$$

$$= W \log_2 \left(1 + \frac{E_b R}{N_0 W}\right) \tag{4.4}$$

where

E_b = energy/bit

R = bit repetition rate

N_0 = noise spectral density in W/Hz

If we take the antilog of Equation (4.4), we can rewrite this equation as

$$\frac{E_b}{N_0} = \frac{W}{R} (2^{C/W} - 1) \tag{4.5}$$

This expression is plotted on the R/W versus E_b/N_0 plane in Figure 4.2. This plane is called the *bandwidth efficiency* plane. Since the ordinate R/W is a measure of

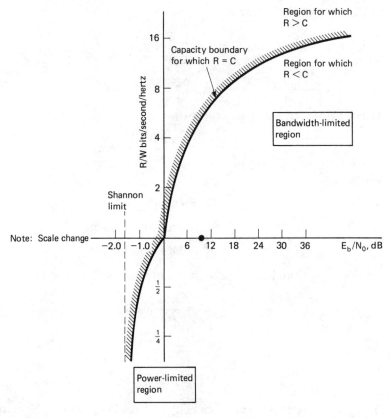

Figure 4.2 Bandwidth efficiency plane. (From B. Sklar, "A Structured Overview of Digital Communications—A Tutorial Review, Part I," *IEEE Communications Magazine,* Aug. 1983. © 1983 IEEE.)

how much data can be transmitted in a specified bandwidth in a given time, it reflects how efficiently the bandwidth resource is used. The abscissa, E_b/N_0, is in decibels. A bit rate $R > C$ cannot be supported by the communication system. For $R < C$, potentially error-free communication is possible.

Using Equation (4.3), we can show that the required E_b/N_0 approaches the required Shannon limit of -1.6 dB as the bandwidth W increases without bound. That is,

$$C = W \log_2 \left(\frac{S}{N} + 1\right) = W \log_2 \left(\frac{S}{N_0 W} + 1\right) \quad (4.6)$$

$$= \frac{S}{N_0} \log_2 \left(\frac{S}{N_0 W} + 1\right)^{N_0 W/S}$$

Next, apply the relation

$$\lim_{x \to 0} (1 + x)^{1/x} = e$$

to Equation (4.6). This results in

$$\lim_{W \to \infty} C = \lim_{W \to \infty} \left[\frac{S}{N_0} \log_2 \left(\frac{S}{N_0} + 1\right)^{N_0 W/S}\right] \quad (4.7)$$

$$= \frac{S}{N_0} \log_2 e$$

The signal power S may be expressed as

$$S = \frac{E_b}{T_b} = E_b \cdot R \quad (4.8)$$

where

$$E_b = \text{energy/bit}$$
$$T_b = \text{time duration/bit}$$
$$R = \text{bit repetition rate}$$

Letting $R = C$ and substituting into Equation (4.7) results in

$$C = \frac{E_b C}{N_0} \log_2 e \quad (4.9)$$

or

$$\frac{E_b}{N_0} = \frac{1}{\log_2 e} = \frac{1}{1.44} = -1.6 \text{ dB}$$

This result is related to the probability of error in Figure 4.3. Note that at the Shannon limit, the probability-of-error curve is discontinuous, going from $P_e = \frac{1}{2}$ to $P_e = 0$. It is not possible to reach the Shannon limit because as the code becomes more complex (increases without bound), the bandwidth requirement and delay become infinite and the complexity of implementing the system increases without bound.

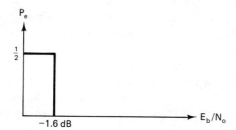

Figure 4.3 Probability of error versus E_b/N_0 at the Shannon limit.

Example 4.1

(a) Let $S/N_0 = 20$ dB and assume a channel bandwidth of 1000 Hz. What is the maximum transmission rate R in bits per second if an arbitrarily complicated system may be used and an arbitrarily small error probability is required?

(b) Repeat (a) if the bandwidth may be as large as desired.

Solution

(a) $C = W \log_2 \left(1 + \dfrac{S}{N}\right)$

$S/N_0 = 20$ dB $\qquad S/(N_0 W) = S/N = 0.2$

$C = R = 1000 \log_2 (1 + 0.2) = 263$ bits/s

(b) Using Equation (4.7), we obtain

$R = \dfrac{S}{N_0} \log_2 e = 1.44(S/N_0) - 1.44(200)$

$ = 288$ bits/s

4.2 CHANNEL ENCODING

Refer to Figure 3.8. Channel encoding refers to the data transformation performed after source encoding but prior to modulation that transforms the source bits into channel bits. Channel encoding is traditionally partitioned into two groups: waveform coding and structured sequences (codes). We will discuss waveform (or signal-design coding) along with baseband codes such as NRZ in Section 4.4. In the following material, we wish to extend our discussion of structured sequences (codes).

We can state the objectives of error-control coding as follows:

1. *Error detection*

   Determine whether a given segment of the received data stream contains errors. Notify the data source or destination of the error.
   Minimize the probability of undetected errors.

2. *Error correction*

   Obtain a reduction in the probability of error for a given value of E_b/N_0
   For a given probability of error, reduce the value of E_b/N_0. The amount of reduction is called the *coding gain* for that probability of error. The concern here is generally with the average probability of error.

Error-detection and error-correction requirements are determined primarily by the type of information being transmitted. For subjective data, such as voice and video, the concern is usually with error correction and average probability of error. The concern with other data types, digital command codes and computer data words, is with error detection and the probability of undetected error. The system throughput is also a matter of concern. The functional block diagram of a transmitter employing coding is shown in Figure 4.4. Referring to this figure, we can make the following observations:

- The symbol rate R_s is always greater than the bit rate R.
- The code rate may be defined as the number of information bits per coded symbol. That is,

$$\text{Code rate} = \frac{R}{R_s} \text{ bits/symbol}$$

4.2.1 Coding Principles

With the foregoing discussion in mind, it is well to examine some of the basic principles involved in implementing codes. The basic problem can be stated as follows: We wish to convert M possible messages into M possible code words. The code words must embody such objectives as error detection, error correction, and security. As an example, consider a message consisting of 3 bits. With 3 bits, there are $2^3 = 8$ possible messages. We wish to substitute a code word for each of the possible messages. As noted previously, to achieve our objectives we employ redundancy. The code words, then, are longer than the corresponding message words from which they are generated.

The error-detection and error-correction capabilities of various redundant codes can be analyzed by first defining the concept of distance between two code words. The distance between any two equal-length binary words is defined to be equal to the number of bit positions by which the two words differ. We can think of distance in terms of multidimensional space. For example, our 3-bit words can be plotted in three-dimensional space (Figure 4.5).

Suppose that we have an elementary message consisting of a 0 or a 1. When the message is 0, we transmit 000. When it is a 1, we transmit 111. Referring to the

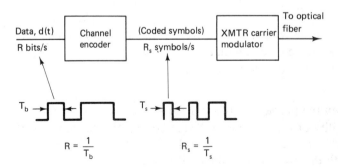

Figure 4.4 Functional block diagram of an optical digital transmitter employing coding.

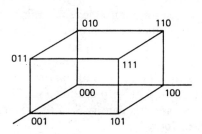

Figure 4.5 Representation of 3-bit words in three-dimensional space.

cube, the distance between the two valid code words representing the two messages is equal to three, because it is necessary to traverse along three of the lines (edges of the cube) to get from one to the other. Now suppose that the message has 3 bits and that all combinations are used. Referring again to Figure 4.5, all the corners of the cube are now used. An error in any one of the 3 bits simply moves us to an adjacent corner, and an error is made. The distance between the transmitted and received word is 1, and an error is made, since the received word corresponds to one of the possible messages. We improve the situation by increasing the distance. For example, the words

$$0000, 0011, 0101, 0110, 1001, 1010, 1100, 1111$$

all have a minimum distance of 2. For these 4-bit code words, we need an abstract figure with 16 points (corners) to represent the four-dimensional code words. It is convenient, then, to think in terms of n-dimensional space.

Using our 4-bit example, suppose that an error is made in one place and that we receive the word 1000. This could have resulted from the words 0000 or 1010. An error is detected, but we cannot correct it. Next consider code words consisting of 01111, 01000, and 10011. The minimum distance is 3. Now suppose that 01000 was transmitted and 01001 was received. The distance to the first code word is 3, the second is 1, and the third is 4. An intelligent receiver can easily be made to decode the received word as 01000. Next suppose that the same message is received as 01011. That is, two errors are made. The received message is now equidistant from 01000 and 10011. An error is detected but cannot be corrected. We say, then, that this code is double-error detecting but only single-error correcting. If the minimum distance between code words is D_{min}, errors involving up to $D_{min} - 1$ can be detected. If D_{min} is an even number, errors up to $(D_{min}/2) - 1$ can be corrected. For D_{min} odd, errors up to $(D_{min})/2$ can be corrected.

The previous ideas are quantitative but do provide valuable insight into the coding process. The reader is referred to a text on digital communications for a treatment of structured codes.

4.3 BANDWIDTH OF DIGITAL DATA

The theorems of Nyquist and Shannon are based strictly on the assumption of band-limited channels. That is, no signal power whatsoever is allowed outside the defined bandwidth. In designing practical systems, then, we are faced with a dilemma.

Strictly band-limited signals are not realizable, since this implies infinite transmission time delay. On the other hand, nonband-limited signals containing energy at arbitrarily high frequencies (wide bandwidth) appear just as unreasonable. In this context, it is easy to see why there is no single universal definition of bandwidth.

Some of the most common definitions of bandwidth are illustrated in Figure 4.6. Other definitions such as -30 or -50 dB down imply even larger bandwidths than those shown. The power spectral density $S(f)$ for a single rectangular pulse of duration τ has the analytical form

$$S(f) = \tau \left[\frac{\sin \pi(f - f_c)\tau}{\pi(f - f_c)\tau} \right]^2 \qquad (4.10)$$

where

f_c = carrier frequency in hertz

and

τ = symbol duration in seconds

This spectral density has the same general $\sin x/x$ appearance as Figure 4.6. It is characteristic of a sequence of random digital data, where we assume that the spectral density averaging time is long relative to the symbol duration τ. Note that the spectral density plot has a main lobe along with smaller symmetrical side lobes. The general shape of this plot is characteristic of that for most digital formats. There are, however, some digital formats that do not have well-defined side lobes.

The bandwidth criteria indicated in Figure 4.6 are as follows:

- BW_1—half-power bandwidth: This is the traditional -3-dB bandwidth, that is, the interval between frequencies at which $S(f)$ has decreased to half-power.
- BW_2—equivalent noise bandwidth (ENBW): This is an equivalent rectangular bandwidth that would contain the total signal power P over all frequencies.

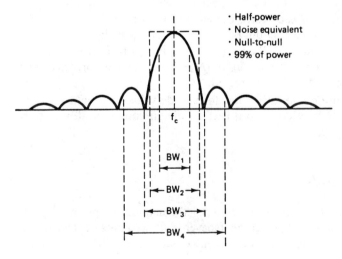

Figure 4.6 Various bandwidth criteria. (From B. Sklar, "A Structured Overview of Digital Communications—A Tutorial Review, Part I," *IEEE Communications Magazine*, Aug. 1983. © 1983 IEEE.)

That is,

$$P = W_N S(f_c) \tag{4.11}$$

where

P = total signal power in watts

W_N = equivalent noise bandwidth in hertz

$S(f_c)$ = the value of $S(f)$ at the center frequency f_c (assumed to be the maximum value over all frequencies)

- BW_3—null-to-null bandwidth: This is the bandwidth of the main lobe. This is a popular measure of bandwidth. Complete generality of this criterion is not possible, however, since some digital formats lack well-defined lobes.
- BW_4—fractional power containment bandwidth: This bandwidth criterion is specified in Section 2.202 of the FCC rules and regulations. Specifically, this definition states that the occupied bandwidth leaves exactly 0.5 percent of the signal power above the upper band limit and exactly 0.5 percent of the signal power below the lower band limit. Thus, 99 percent of the signal power is contained inside.

According to Nyquist's theorem, it is theoretically possible to transmit a pulse train of repetition frequency f_r over a channel bandwidth $\frac{1}{2}f_r$, which is the Nyquist bandwidth. The received pulse is of the form $\sin x/x$, with zeros of the tails occurring at multiples of a time slot. This theoretically leads to no ISI but to an eye that requires perfect clock stability and zero clock sample width.

Example 4.2

It is desired to transmit speech over a PCM channel with 8-bit accuracy. Assume the speech is baseband-limited to 4 kHz. Determine the minimum bandwidth (BW_{min}) required for the PCM signal.

Solution

Bit rate = 2×4000 = 64,000 bits/s

$BW_{min} = \frac{1}{\tau} = \frac{1}{2}f_r = 32,000$ Hz

Example 4.7 illustrates the calculation of PCM bandwidth needed under the ideal conditions postulated by Nyquist. A more practical number is arrived at by considering the frequency spectrum of a single pulse (Figure 4.7). For design purposes, the bandwidth is often taken to be the distance to the first zero crossing. That is,

$$BW = \frac{1}{\tau} \text{ Hz} \tag{4.12}$$

where τ is the width of the pulse. Now our PCM wave train is composed of a sequence of ones and zeros. Regardless of the sequence, the bandwidth can be no greater than that required for a single pulse, and the bandwidth is as expressed by

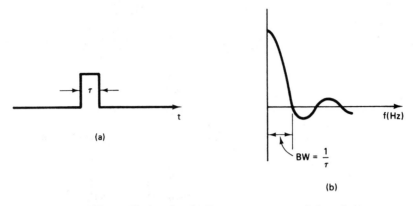

Figure 4.7 (a) Single pulse. (b) Frequency spectrum of the pulse.

Equation (4.12). Let us suppose that we sample an analog signal at the Nyquist rate. The time T between pulses is then given by

$$T = \frac{1}{2f_m} \tag{4.13}$$

where f_m is the highest frequency in the signal. Extending this to n signals (multiplexing), the width of the sampled pulse will be

$$\tau_n = \frac{T}{n} = \frac{1}{2nf_m} \tag{4.14}$$

If we wish to digitize each pulse, each code word can last only T/n seconds. For an m-bit PCM word, the width of each bit pulse is then T/mn. Using Equation (4.12), the PCM bandwidth, BW_{PCM}, is given by

$$BW_{PCM} = \frac{1}{\tau} = \frac{1(mn)}{T} \text{ Hz} \tag{4.15}$$

$$= (m)\left(\frac{\omega_m}{\pi}\right) \text{ Hz}$$

For the binary process, the number of levels L in the quantization process is given by

$$L = 2^m \tag{4.16}$$

Solving for m and substituting the result in Equation (4.15) results in

$$BW_{PCM} = \left(\frac{n\omega_m}{\pi}\right)(\log_2 L) \text{ Hz} \tag{4.17}$$

Example 4.3

A PCM system transmits 24 time-multiplexed voice signals, each bandlimited to 8 kHz. What bandwidth is required if an 8-bit PCM word is used?

Solution

$$BW = (2nf_m)(m)$$
$$= (2 \times 24 \times 8000)(8)$$
$$= 3.072 \text{ MHz}$$

4.4 DIGITAL SIGNALING TECHNIQUES

Techniques for digital signaling (encoding) were introduced briefly in Chapter 3 (see Section 3.2). The emergence of local networks and the ongoing evolution of the public telecommunications network to digital service has led to an increased interest in digital signaling techniques. The precise meaning of this terminology is illustrated in Figure 4.8. Digital data is generated by the source and is typically represented as discrete voltages—one voltage level for binary and another for binary 1. This format is referred to as NRZ-L (nonreturn to zero-level). These pulses could be transmitted directly. It is often desirable, however, to encode the data in such a way as to improve performance.

Figure 4.8 Digital signaling with digital data.

We see from Figure 4.8 the relation between digital data and digital signals. A digital signal is a sequence of discrete, discontinuous voltage pulses. Each pulse is a signal element. The binary data is transmitted by encoding each data bit into signal elements. Now, if the signal elements all have the same algebraic sign, that is, all positive or all negative, the signal is said to be *unipolar*. Conversely, however, if one logic state is represented by a positive voltage level, and the other by a negative voltage level, the signal is said to be *polar*. The *data rate* of a signal is the rate in bits per second at which the data is transmitted. For a data rate of R bits/s, the bit duration is $1/R$. The *modulation rate* is the rate at which the signal level is changed. We shall see that this depends on the nature of the digital encoding. The modulation rate is commonly expressed in *bauds*, which are signal elements per second. For historical reasons, we also encounter the terms *mark* and *space*, which refer to the binary digits 1 and 0, respectively.

In a fiber optic system, the digital signal is used to modulate an LED or laser diode. The received signal is recovered by the photodetector receiver. At the receiver, two important tasks are involved in interpreting the digital signal. First, the receiver must know the timing of each bit. That is, the receiver must know when a bit begins and ends. Second, the receiver must determine the signal level for each bit position— high for logic 1 and low for logic 0.

A number of factors are involved in determining the success of the receiver interpreting the incoming signal. Among these are the signal-to-noise ratio (SNR), the data rate, R, and the bandwidth of the signal. We can make the following generalized statements:

- An increase in data rate increases the bit error rate.
- An increase in SNR decreases the error rate.
- An increase in bandwidth allows an increase in data rate.

In addition to these items, there is another factor that can be used to improve performance. This is an encoding scheme that maps data bits to signal elements. A number of the more common ones are itemized in Table 4.1 and are illustrated in Figure 4.9. These signal encoding formats can be categorized as follows:

- Nonreturn to zero (NRZ)
- Return to zero (RZ)
- Biphase
- Delay modulation
- Multilevel binary

Before we discuss these techniques, it is of interest to consider ways of evaluating or comparing the techniques. Some evaluation techniques that have been proposed are

TABLE 4.1   DEFINITION OF DIGITAL SIGNAL ENCODING FORMATS

Nonreturn to zero-level (NRZ-L)
  1 = high level
  0 = low level
Nonreturn to zero-mark (NRZ-M)
  1 = transition at beginning of interval
  0 = no transition
Nonreturn to zero-space (NRZ-S)
  1 = pulse in first half of bit interval
  0 = no pulse
Biphase-level (Manchester)
  1 = transition from high to low in middle of interval
  0 = transition from low to high in middle of interval
Biphase-mark
  Always a transition at beginning of interval
  1 = no transition in middle of interval
  0 = transition in middle of interval
Differential Manchester
  1 = no transition in middle of interval
  0 = transition at beginning of interval
Delay modulation (Miller)
  1 = transition in middle of interval
  0 = no transition if followed by 1
  Transition at end of interval if followed by 0
Bipolar
  1 = pulse in first half of bit level, alternating polarity pulse to pulse
  0 = no pulse

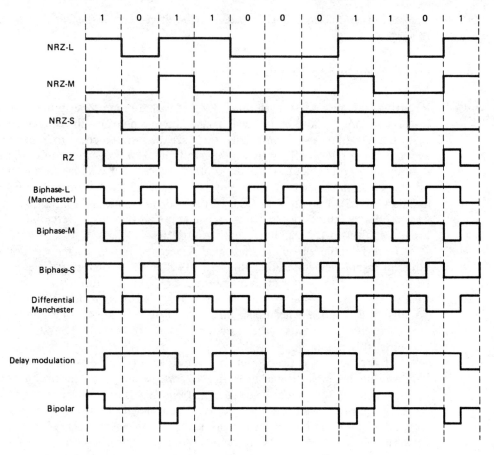

Figure 4.9 Digital signal-encoding formats. (From W. Stallings, "Digital Signaling Techniques," *IEEE Communications Magazine,* Dec. 1984. © 1984 IEEE.)

- *Signal frequency spectrum.* Less bandwidth is required if there is no high frequency content. Conversely, if there is no DC component, AC coupling can be employed.
- *Signal synchronization capability.* We previously mentioned the need to determine the beginning and end of each bit position. It is helpful if information is contained in the signal frequency spectrum that will permit the synchronization of a clock in the receiver. This is sometimes referred to as a *self-clocking capability*.
- *Signal interference and noise immunity.* Some codes exhibit superior performance in the presence of noise. This usually is reflected by a smaller BER rate for the same SNR.
- *Error-detection capability.* Some signaling schemes inherently possess this feature.
- *Cost and complexity.* Factors that must be considered in any design.

The various signal encoding formats can be discussed on the basis of the above factors.

4.4.1 Nonreturn to Zero (NRZ)

Referring to Figure 4.9, we see that NRZ codes share the property that the voltage level is constant during a bit interval. That is, there is no transition (no return to zero voltage level). In terms of complexity, these are the simplest codes to implement, the simplest being NRZ-L. This code is generally used to generate or interpret digital data by data processing terminals and other devices. If we desire to transmit with another code, this code is typically generated from the NRZ-L signal (see Figure 4.9 and Section 4.5). Two other versions of this code are mark and space, that is, NRZ-M and NRZ-S. These codes have the advantage that they are differential codes and are less subject to DC line-bias voltage-shift effects.

In differential encoding, the signal is decoded by comparing the polarity of adjacent signal elements rather than the absolute value of a signal element. Two benefits accrued from this: (1) It may be more reliable to detect a transition in noise than to compare a value to a threshold, and (2) it is easy to lose the sense of the polarity of a signal. For example, on a multidrop twisted-pair line, the accidental reversal of the lines results in the inversion of 1s and 0s for NRZ-L. With differential encoding, this cannot happen.

NRZ codes are easiest to engineer and also make the most efficient use of bandwidth. This is illustrated in Figure 4.7, where the power spectral densities of various encoding schemes are plotted. The frequency is normalized to the data rate. Note that most of the energy in an NRZ signal lies between DC and half the bit rate. That is, with a bit rate of 9600 bits/s, most of the signal energy is concentrated between DC and 4800 Hz. With the advantages mentioned we may well ask ourselves why we consider encoding schemes. Consider the disadvantages. The primary limitations of NRZ signals are the presence of a DC component and the lack of a self-synchronization capability. For example, with a long string of 1s for NRZ-L or NRZ-S, the output is a constant voltage. Based on the signal alone, any drift between the transmitter and receiver cannot be corrected. These limitations make these codes unattractive for signal-transmission applications. However, because of their simplicity and relatively low frequency-response characteristics, NRZ codes are commonly used for digital magnetic recording and for asynchronous coding, where synchronization is provided by the occurrence of the start pulses.

4.4.2 Return to Zero (RZ)

The RZ format actually provides no improvement over the previously discussed NRZ techniques. Referring to Figure 4.4, we note a distinction between data rate and modulation rate. The data rate is $1/t_B$, where t_B is the bit repetition period. The modulation rate is the average number of transitions that occur per second. Thus, for RZ, the maximum modulation rate R is given by

$$R = \frac{2}{t_B} \text{ baud} \qquad (4.18)$$

This occurs during a string of 1s. The minimum naturally occurs during a string of 0s. Because the modulation rate is twice that for NRZ, the bandwidth of the signal is also twice as large. The limitations of NRZ apply to RZ as well. That is, there is a DC component and no synchronization (with a string of zeros). RZ is used in some applications because of its simplicity but is not a preferred choice for most applications.

4.4.3 Biphase

A number of biphase encoding formats exist. Actually, the term biphase encompasses biphase-L (Manchester), biphase-M, biphase-S, and differential Manchester. These schemes are devised to overcome the limitations of NRZ and RZ signal-encoding techniques. A common characteristic of biphase schemes is that they all require at least one transition per bit period. There may be as many as two transitions per bit period (see Table 4.2). We see then that the modulation rate R for biphase is twice that for NRZ with a corresponding greater bandwidth.

TABLE 4.2 SIGNAL TRANSITION RATE

Code	Minimum	101010 . . .	Maximum
NRZ-L	0 (all 0s or 1s)	1.0	1.0 (1010 . . .)
NRZ-M	0 (all 0s)	0.5	1.0 (all 1s)
NRZ-S	0 (all 0s)	0.5	1.0 (all 0s)
RZ	0 (all 0s)	1.0	2.0 (all 1s)
Manchester	1.0 (1010 . . .)	1.0	2.0 (all 0s or 1s)
Biphase-M	1.0 (all 0s)	1.5	2.0 (all 1s)
Biphase-S	1.0 (all 0s)	1.5	2.0 (all 0s)
Differential Manchester	1.0 (all 1s)	1.5	2.0 (all 0s)
Delay modulation	0.5 (1010 . . .)	0.5	1.0 (all 0s or 1s)
Bipolar	0 (all 0s)	1.0	2.0 (all 1s)

The question then, again, is why we use the biphase schemes. There are several advantages:

- *Synchronization.* Referring to Figure 4.6, we see that there is a predictable transition during each bit period. The receiver can synchronize on this transition. Specifically, for Manchester and differential Manchester, there is always a transition in the middle of the bit interval. For biphase-M and biphase-S, there is always a transition at the beginning of a bit time. Biphase codes, then, are known as *self-clocking codes.*
- *No DC component.* The guaranteed transition ensures that the average DC level is zero.

- *Error detection.* The absence of an expected transition can be used to detect errors. Noise would have to invert the signal both before and after the expected transition to cause an undetected error.

Referring to Figure 4.10, note that most of the energy in biphase codes is between one-half and one times the bit rate. There is no DC component. All the biphase codes are differential with the exception of Manchester.

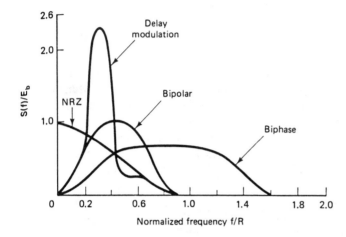

Figure 4.10 Spectral density of various digital signal-encoding schemes. (From W. Stallings, "Digital Signaling Techniques," *IEEE Communications Magazine*, Dec. 1984. © 1984 IEEE.)

NRZ is still the most widely used code in data communication systems. Biphase codes are also popular techniques for data transmission, and the Manchester code is gaining ground rapidly. This code is popular in magnetic tape recording and is used as an input signal for fiber-optic modulation systems. Local area network (LAN) standards specify both Manchester and differential Manchester. In fact, the Manchester code has been specified for the IEEE 802.3 LAN standard for baseband coaxial cable using carrier sense multiple access/collision detection (CSMA/CD), that is, Ethernet. Differential Manchester has been specified for the IEEE 802.5 standard for the token ring LAN, using either baseband coaxial cable or twisted pair. Differential Manchester is preferred for twisted pair because it uses differential encoding.

4.4.4 Delay Modulation

Delay modulation, also known as *Miller encoding,* is an interesting alternative to the biphase technique. With this technique there is at least one transition per two bit times. There is never more than one transition per bit. We see, then, that Miller coding requires less bandwidth than biphase and has a lower modulation rate. It also possesses some synchronization capability. Referring to Figure 4.10, we see that the bandwidth for Miller is considerably less than for either NRZ or biphase. However, the figure is misleading. For worst-case bit patterns, Miller can have a bandwidth greater than NRZ along with a significant DC component.

The BER for several digital encoding schemes is shown in Figure 4.11. Note that Manchester and NRZ exhibit identical performance with an improvement over Miller of 3 dB. The derivation of formulas leading to these curves is based on the probability of error in white noise. Without entering into this analysis we can intuitively see the reason for the results. At the receiver a decision is made based on whether a received bit interval contains 1 or 0. For both NRZ and Manchester, there are only two elementary pulse waveforms from which to choose. This statement also applies to RZ, but RZ makes inefficient use of signal time. There is either a half-pulse or no pulse. The decision is even more difficult for the Miller code, because it makes use of four elementary codes.

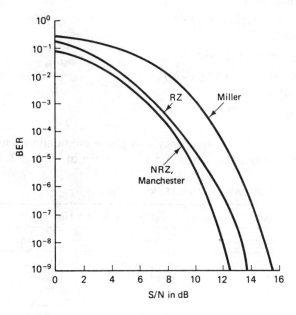

Figure 4.11 Theoretical BER for various digital encoding schemes. (From W. Stallings, "Digital Signaling Techniques," *IEEE Communications Magazine,* Dec. 1984. © 1984 IEEE.)

4.4.5 Multilevel Binary

Multilevel binary uses more than two signal levels. An example of this is bipolar (Figure 4.9). It is widely used by AT&T for T1-PCM carriers (see Chapter 5). From Figure 4.10, we see that it has a bandwidth centered on one-half the bit rate. In terms of characteristics, there is no DC component, and a lack of synchronization capability. Some error-detection capability does exist, since successive 1s must have opposite signs.

4.5 DATA ENCODING FOR FIBER OPTICS

Encoding data often permits a long string of 1s or 0s to be transmitted. This can result in problems in a high-speed fiber-optic link. Since light pulses are unipolar (either on or off), the receiver must be AC-coupled to eliminate DC-drift errors.

Thus the data must contain a transition during each bit period. Nonreturn to zero (NRZ) can prove to be unsuitable for high data rates. Manchester coding (biphase) can be used to transmit data at better than 1 Mbits/s. The disadvantage of biphase, however, is that the required bandwidth is twice that for NRZ. This effectively halves the bit rate.

More recently developed codes are better suited than Manchester and in some cases often improve bandwidth. The most important codes suitable for fiber optics use are

Frequency-shift code (FSC)
Frequency-shift keying (FSK)
Modified frequency-shift keying (MFSK)
Phase-shift keying (PSK)
Modified phase-shift keying (MPSK)
Delayed modified phase-shift keying (DMPSK)
Phase-delay keying (PDK)
Modified phase-delay code (MPDC)

These codes are based on digital frequency and phase modulation schemes. The characteristics of these data-transmission codes are shown in Table 4.3.

TABLE 4.3 CHARACTERISTICS OF DATA-TRANSMISSION CODES

Code	Bit Duration[a] Min.	Bit Duration[a] Max.	Patterns to Achieve High Speed[b] Worst	Patterns to Achieve High Speed[b] Best
NRZ	t	—	Alternate 1s and 0s	All 1s and 0s
Manchester	$t/2$	t	All 1s or all 0s	Alternate 1s and 0s
FSC	$t/2$	t	All 1s	All 0s
PSK	$t/2$	$3t/2$	All 1s	Alternate 1s and 0s
MPSK	t	—	All 1s or all 0s	Alternate 1s and 0s
DMPSK	$t/2$	t	Alternate 1s and 0s	All 1s or all 0s
PDK(90)	$t/2$	$3t/2$	The first 1 after 0	The first 0 after 1
MPDC	t	$2t$	All 1s or all 0s	The first 0 after 1

[a] For easy comparison, the tabulated codes each use clocks with periods t.
[b] Data-coding schemes that feature wide spacing between transitions cut the chance of decoding errors.

The Manchester code polarity-reverses in each bit period, regardless of the information. This transition occurs during the split-phase period at the bit's center. Positive-to-negative reversal indicates a binary 1 and negative-to-positive indicates a binary 0. To generate Manchester from NRZ, two clock phases are needed (Figure 4.12). The encoded logic is shown in Figure 4.12(b). Based on an RS flip-flop, the set and reset logic functions are defined as

$$\text{Set} = A \cdot \text{data} + B \cdot \overline{\text{data}}$$

$$\text{Reset} = A \cdot \overline{\text{data}} + B \cdot \text{data}$$

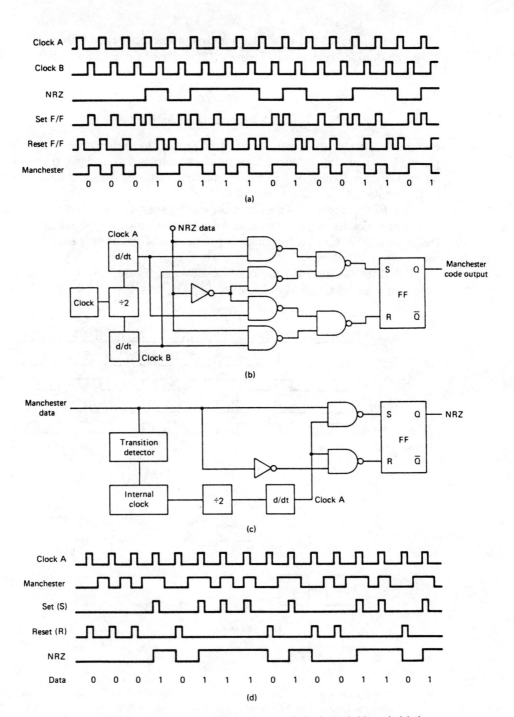

Figure 4.12 Manchester-coded data reverse polarity in each bit period independently of the bit value. A positive-to-negative reversal at the bit's center signifies "one," while negative-to-positive means "zero." (b) While you need a two-phase clock to encode NRZ data into Manchester, (c) a single-phase clock is sufficient for decoding Manchester waveforms into NRZ. For synchronization, the decoder retimes its clock to received-signal transitions. (Reprinted with permission from *Electronic Design*, Oct. 25, 1978. Copyright 1978, Penton Publishing.)

The output of the RS flip-flop is the required Manchester code with an NRZ input.

The decoder is shown in Figure 4.12(c). Based on an RS flip-flop, the logic functions are

$$\text{Set} = A \cdot \text{data}$$

$$\text{Reset} = A \cdot \overline{\text{data}}$$

For the decoder to work it must be synchronized to the encoder. This is accomplished by having the decoder adjust its clock, using received signal transition. The internal clock is divided by 2 to regenerate clock phase A. A similar scheme for FSC code is shown in Figures 4.13 and 4.14.

An FSK encoder circuit is illustrated in Figure 4.15. In this particular circuit, two frequencies are derived which are keyed to the data. Since the phase of the two frequencies is constant, a transition is not needed at the leading edge of each data bit.

NRZ to PSK may be produced as shown in Figure 4.16. If we compare the waveforms for PSK and Manchester coding (Figures 4.16 and 4.12), we note that

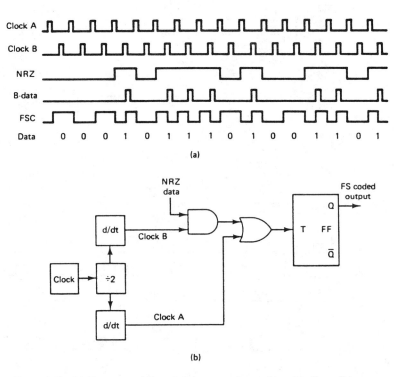

Figure 4.13 (a) Frequency shift coded data waveforms show that "ones" have an extra level at the data bit's center, while "zeros" have only the single level change at the data bit's start. (b) The encoder uses two streams of clock pulses—A and B. These pulses are 180 degrees apart and can be derived from the clock. (Reprinted with permission from *Electronic Design,* Oct. 25, 1978. Copyright 1978, Penton Publishing.)

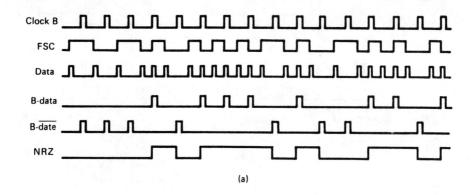

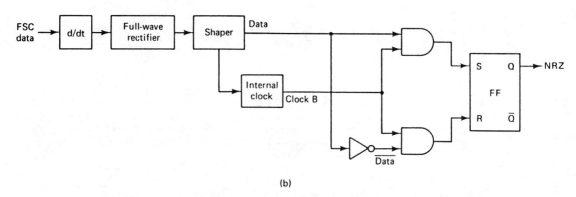

Figure 4.14 (a) In frequency shift code, the direction of the received-data's transition relative to the clock is not a factor. (b) Only one clock phase is required to reconstruct the original NRZ data from the received FS code. (Reprinted with permission from *Electronic Design*, Oct. 25, 1978. Copyright 1978, Penton Publishing.)

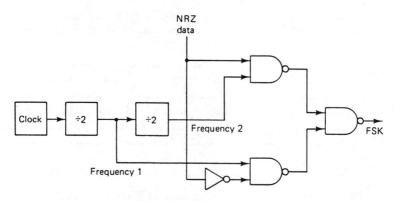

Figure 4.15 FSK encoder circuit generates its two clocks by dividing down from the higher frequency. (Reprinted with permission from *Electronic Design,* Oct. 25, 1978. Copyright 1978, Penton Publishing.)

Sec. 4.5   Data Encoding for Fiber Optics

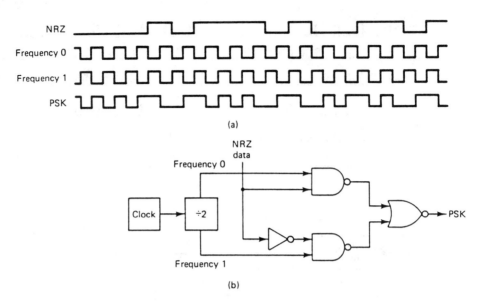

Figure 4.16 (a) Phase shift keying waveforms show a carrier shifted by 180 deg at each data change. The carrier phase always represents the bit being transmitted at any particular time. (b) The encoder uses two identical-frequency square-wave clocks that are phased 180 deg apart to produce waveforms identical to those in Manchester coding. (Reprinted with permission from *Electronic Design,* Oct. 25, 1978. Copyright 1978, Penton Publishing.)

the two are identical. In essence, then, Manchester encoding is just a form of PSK coding. With basic PSK (Figure 4.16) a full square-wave cycle (positive and negative transition) is needed for each bit. The value of the bit will be indicated by the transition's direction at the bit's center. Thus the transmitted signal frequency (PSK) is twice that the clock used to shift the original NRZ data.

The basic PSK coding can be modified so that the 180-deg phase shifts correspond to data-state transitions rather than to the data. This permits clocking at half the basic PSK frequency. Typical waveforms for MPSK are shown in Figure 4.17. In order to produce these waveforms, the clock must be in phase with the data.

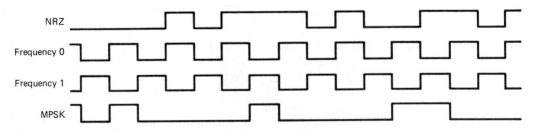

Figure 4.17 By modifying basic phase-shift-keying code so that the 180-deg phase shifts correspond to data transitions rather than to data, you can clock the resulting MPSK code at half the frequency required by phase-shift coding. (Reprinted with permission from *Electronic Design,* Oct. 25, 1978. Copyright 1978, Penton Publishing.)

MPSK results in the same data transmission rates as the original NRZ code and at the same bandwidth. However, there is a drawback. DC levels similar to NRZ occur. Note from Figure 4.17 that no changes occur in the MPSK waveform when alternating ones or zeros are transmitted. MPSK can be produced with a circuit like the one in Figure 4.16(b), except that a slower clock is used. Decoding is accomplished simply by detecting whether or not a transition occurs at the leading edge of each bit. The absence of a transition means that the NRZ data has changed.

An improvement of MPSK coding can be effected by shifting the modulation frequencies 90 degrees relative to the data. In this case the data and the clock are no longer in phase. The resulting delayed MPSK (DMPSK) produces clock transitions at data-bit centers. The advantages are bit-center data-level transitions required by the PSK code along with the lower frequency clock required by MPSK. Techniques for producing MPSK from NRZ are illustrated in Figure 4.18. Note that this encoder is the PSK encoder with the exception that the timing generator is more complex. In DMPSK, the widest pattern spacing occurs when transmitting all ones or all zeros. The decoding circuit is illustrated in Figure 4.19. This circuit simply tests for transitions at the start of each bit's time slot without regard to direction. After detecting the DMPSK waveform, the decoder full-wave rectifies and shapes the transitions into pulses. When a transition pulse coincides with a clock pulse (A), the NRZ output level must shift. The transition pulses and clock toggles the output flip-flop.

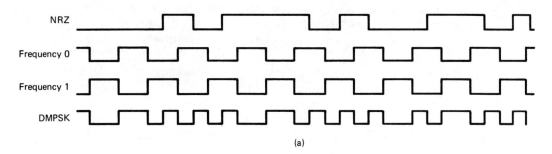

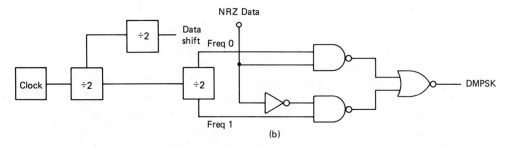

Figure 4.18 (a) Shifting modulation clocks 90 deg from the MPSK scheme gives you data-level transitions at the bit centers and another coding variety-delayed DMPSK. (b) Only a more complex timing circuit distinguishes the DMPSK encoder circuit, but DMPSK gives you bit-center transitions and MPSK's low clock rate. (Reprinted with permission from *Electronic Design*, Oct. 25, 1978. Copyright 1978, Penton Publishing.)

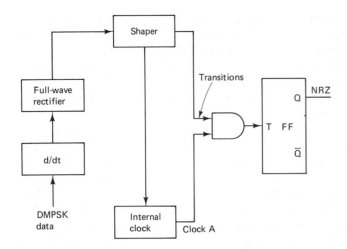

Figure 4.19 When DMPSK is decoded into NRZ, only the transitions at the start of each bit's time slot matter. When a transition pulse and clock-pulse A coincide, the output toggles. (Reprinted with permission from *Electronic Design,* Oct. 25, 1978. Copyright 1978, Penton Publishing.)

The generation of FSK is illustrated in Figure 4.15. FSK produces better results than PSK, even with the lower frequencies. Other PSK variations exist, however, which also have the relatively narrow bandwidth of FSK along with more useful patterns. Using clocks with the same frequency as in MPSK, one in phase with the data and the other delayed by either 90 deg or 270 deg, results in two phase-delay-keying codes with the waveforms illustrated in Figure 4.20. Note from this figure that two additional PDK waveforms are possible. They are, however, simply the complements of the 90-deg and 270-deg codes. The worst pattern in PDK (90 deg) results when a one follows a zero. The best pattern results when the first zero follows a one. PDK (90 deg) may be decoded by using the same circuit as FSC and FSK (Figure 4.14). In the decoding process for PDK, the clock transition directions at clock-time B and clock-time A are not important. Only the presence or absence of data pulses at clock-time B matters. The presence of a data pulse means that the NRZ output changes from zero to one. The absence of a data pulse means a change from one to zero. Based on an RS flip-flop, the logic functions are

$$\text{Set} = B \cdot \text{data}$$

$$\text{Reset} = B \cdot \overline{\text{data}}$$

Almost identical results are obtained when the PDK (270 deg) scheme is decoded. The only difference is that at clock-time A the NRA output changes to one if a pulse is absent or to zero if a pulse is present.

Figure 4.20(a) illustrates that the worst PDK (90 deg) pattern occurs when the first one comes after a zero. This degraded pattern results from a short-duration data transition pulse, whose width equals half the clock spacing. This short pulse does not occur for each one. It occurs only for ones following a zero. Referring to the waveforms for PDK (90 deg) and NRZ in Figure 4.20, we see that these short data transition pulses are caused by a transition from zero at time A to one at time B. Therefore, we can regard the edge at time A, not as the leading edge of pulses representing ones, but more properly as the zero's trailing edge.

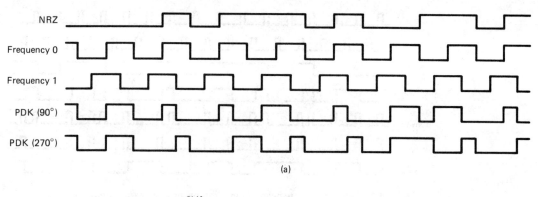

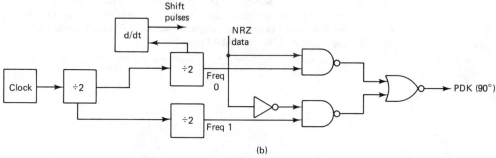

Figure 4.20 (a) Clocks for phase-delay-keying code are the same as those for PSK, except that they are delayed from each other by either 90 deg or 270 deg. (b) The PDK encoder differs from the DMPSK circuit in that one of its clocks is in phase with the incoming NRZ data. The complements of the 90-deg and 270-deg codes are also possible in PDK. (Reprinted with permission from *Electronic Design,* Oct. 25, 1978. Copyright 1978, Penton Publishing.)

Now, if we use all of the other transitions at times A and B, we should be able to ignore the leading edge of the narrow pulse at time A. Figure 4.21 illustrates the basic waveforms for such a scheme-modifed phase-delay (MDP) code. This figure illustrates an NRZ waveform, its corresponding PDK (90 deg) waveform, and the PDK code's transition pulses. In this stream of transition pulses, the first pulses representing a one coming after a zero at time A are removed. The resulting phase train generates an MPD code, which has a level transition for each pulse left in the transition pulse train. Because the transitions do not follow a fixed frequency, an MPD code is not just a simple keying code, even though it is based on phase-delay keying. An MPD code has bit-center transitions for ones and bit-leading transitions for zeros with the exception of the first zero after a one. With the exception of the first zero after a one, the *A* clock starts zeros, while the *B* clock starts ones.

This code yields the best results without increasing the frequency above the basic clock frequency. The best pattern is generated for the first zero after a one. This pattern will repeat when alternate ones and zeros are transmitted. The worst pattern occurs when all ones or all zeros are transmitted. Thus, the data received is least susceptible to error when the transmitted pattern is alternate ones and zeros.

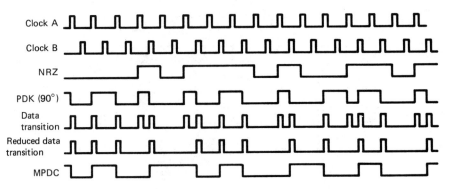

Figure 4.21 Modified phase-delay coding features bit-center transitions for zeros. In this rather complex scheme, zeros start at the A clock and "ones" start at the B, except for the first "one." (Reprinted with permission from *Electronic Design,* Oct. 25, 1978. Copyright 1978, Penton Publishing.)

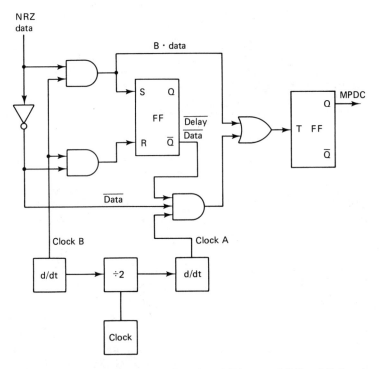

Figure 4.22 The MPDC encoder delays the NRZ input a full "one's" duration and thus inhibits the first zero after each "one." MPDC coding generates its best pattern for the first zero after a "one." The worst pattern in this nonfixed frequency system comes with all "ones" or all zeros. (Reprinted with permission from *Electronic Design,* Oct. 25, 1978. Copyright 1978, Penton Publishing.)

These patterns and those similar to it should be reserved for crucial synchronization codes. It is obvious that we need to avoid data streams consisting of all ones or all zeros.

When NRZ data is encoded into MDP code, it is necessary to inhibit the first zero after each one. This can be accomplished by using the circuit illustrated in Figure 4.22. This circuit delays the NRZ-coded input by the full bit period of a one. In order to avoid voltage glitches at time A (when the zero is strobed), delay the data by half of a bit period. Next, compare the original NRZ with the delayed NRZ, and look only at those A pulses that occur when both NRZ streams are at zero. The output flip-flop is toggled with each B pulse that occurs when both NRZ streams are at zero. The output flip-flop is toggled with each B pulse for which the original data are ones and with each A pulse for which the original and delayed data are zeros. Decoding can be accomplished by using the circuit illustrated in Figure 4.14. Thus, after encoding the NRZ data into MPDC or FSC, we can transmit either over the same channel to the same decoder.

REFERENCES

4.1 JOHANNES, V. I. 1984. Improving on bit error rate. *IEEE Communications Magazine* 22:18–20.

4.2 STALLINGS, W. 1984. Digital signaling techniques. *IEEE Communications Magazine* 22:21–25.

4.3 KILLEN, H. B. 1988. *Digital Communications with Fiber Optics and Satellite Applications.* Englewood Cliffs, N.J.: Prentice Hall.

4.4 MORRIS, D. J. 1978. Code your fiber-optic data for speed without losing circuit simplicity. *Electronics Design* 22:84–91.

4.5 MASSEY, J. L. 1984. Information theory: the Copernican system of communications. *IEEE Communications Magazine* 22:26–28.

PROBLEMS

4.1. It can be shown that the bit error rate of an MPSK digital communication system is given by

$$\text{BER} \log_2 n \, [\text{erfc}(x)]$$

where

$$n = \text{number of phases}$$

$$x = [\sin \pi/n (\log_2 n) E_b/N_0]^{1/2}$$

and the complementary error function is given by

$$\text{erfc}(x) = \frac{1}{\sqrt{2\pi}} \int_x^\infty e^{-y^2/2} \, dy$$

Integrate by parts to establish the bounds

$$\text{erfc}(x) < \frac{1}{\sqrt{2\pi}} \exp[-x^2/2], \qquad x > 0$$

$$\text{erfc}(x) > \frac{1}{\sqrt{2\pi}} (1 - 1/x^2) \exp[-x^2/2], \qquad x > 0$$

4.2. Plot the capacity vs bandwidth of a channel with additive white Gaussian noise of spectral density N_0 and average signal power P. What is the asymptotic value of the capacity?

4.3. Received data pulses have a 30 percent average duty cycle and a 1-ns pulse width. If the average received power is -69 dBm, find:
(a) Energy/bit
(b) Number of photons in one received light pulse if the transmitter is operating at 1.55 μm
(c) What is the shortest wavelength detector that would respond to the received pulses?

4.4. Calculate the capacity of a low-pass channel with a bandwidth of 3400 Hz and an SNR at the output of 1000. The channel noise is Gaussian and white.

4.5. An ideal low-pass channel of W Hz bandwidth with additive Gaussian white noise is used for transmitting digital information. Plot C/W vs SNR in decibels for this system.

4.6. (a) Calculate the capacity of a Gaussian channel with a bandwidth of 1 MHz and an SNR of 30 dB.
(b) How long will it take to transmit two million ASCII characters over this channel? In ASCII, each character is coded as 8 bits. Ignore stop and start bits.

4.7. An analog signal with 4-KHz bandwidth is sampled at 2.5 times the Nyquist rate. Each sample is quantized to 256 equally likely levels.
(a) Can the output of this system be transmitted without errors over a Gaussian channel with 50-KHz bandwidth and an SNR of 23 dB?
(b) What will be the bandwidth requirements of an analog channel for transmitting the output of the source without errors if the SNR is 10 dB?

4.8. What two properties do Manchester and biphase codes possess that are lacking in R codes?

4.9. The received C/N_0 ratio in a digital link is 86.5 dB·Hz and the data rate is 50 Mbits/s. Calculate the E_b/N_0 ratio.

4.10. For problem 4.9, calculate the bandwidth if the transmission is by **(a)** RZ, **(b)** NRZ.

5

Digital Video Transmission in Optical Fiber Networks

5.0 INTRODUCTION

Recent advances in analog fiber systems are permitting lightwave-transmitted analog video signals. Techniques for accomplishing this were discussed in Chapter 2. In the long run, however, proponents of both analog and digital systems agree that the format is likely to be digital. At any rate, the transmission of high-quality video over fiber optic lines is proving to be an irresistible lure. The choice is obvious because of the wide bandwidth required by video. Other significant advantages are

- Long span lengths without amplifiers or repeaters
- Low noise level
- Transmission unaffected by weather
- Resistance to interference, consistent performance with high reliability
- Advanced networking, including software-controlled switching, time division multiplexing, and drop and insert capability. This allows for mixing voice, data, and video transparently at the network level.
- Sophisticated encryption, when necessary

In considering a new design for transmitting video, it is prudent to "future-proof" the network. Thus, the exclusive use of single-mode fiber must be considered because of the extremely large bandwidth. In addition, because of technological advances, single-mode fiber is very cost-effective today. While light emitting diodes and analog modulation techniques can be used effectively with multimode fiber, the small core size of single-mode fibers creates technical problems with coupling power from the source to the fiber. In order to overcome this problem, the source must have a large facet of power that is highly directional. We know from Chapter 2 that these requirements are met by a laser. Refer to Figure 1.22. For wide bandwidths, direct analog modulation of a laser is difficult. Thus digital techniques (NRZ,RZ baseband modulation) or pseudo-digital techniques such as pulse-width modulation (PWM) or pulse-frequency modulation (PFM) must be used. For these reasons, single-mode fiber with a laser as the optical source and digital modulation is the preferred choice.

5.1 FACTORS AFFECTING DIGITAL VIDEO

In previous chapters, we discussed technical details related to designing fiber optic systems. The transmission of video poses particular challenges, not all of which are technical. Recent advances in technology, especially in CMOS, have led to a sizable reduction in cost, power consumption, and the size of digital circuitry. Further reductions in cost will come with large-scale integration. Before this can happen, however, the industry must commit itself to standards. The scenario affecting the transmission of digital video is illustrated in Figure 5.1. Today, various standards such as SONET (Synchronous Optical Network), ISDN (Integrated Services Digital Network), and FDDI (Fiber Distributed Data Interface) are under consideration. Increased deployment of optical fiber for video transmission is creating a critical need for standards.

In the United States, the use of DS3 (Data Signal Level 3—see Figure 5.2) is justified as a standard rate for good-quality video for several reasons. First, to obtain good-quality video with no perceptible degradation, a transmission capacity of

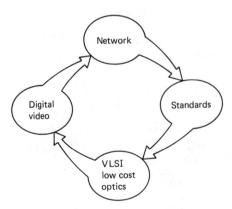

Figure 5.1 Standards development, the advancement of low-cost VLSI optics, and the increased use of fiber optic networks all affect the growth of digital video systems.

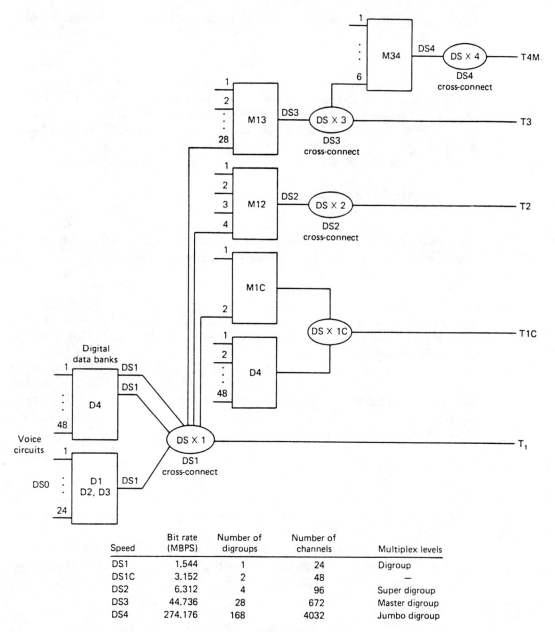

Figure 5.2 Digital transmission facility hierarchy. (From M. F. Slana and H. R. Lehman, "Data Communications Using the Telecommunication Network," *Computer,* May 1981. © 1981 IEEE.)

at least 30 to 45 Mbits/s is required. Second, there is an abundance of DS3 capacity, due primarily to the several high-capacity fiber-optic systems that are in place (565 Mbits/s and 1.13 Gbits/s) for long-haul and interoffice applications. In addition,

Sec. 5.1   Factors Affecting Digital Video   145 there also exist digital radio DS3 networks that can serve areas where fiber installation is expensive or difficult. The availability of 18-GHz subscriber radio provides an easy DS3 facility on an on-demand basis. Networking of DS3s is simplified by DS3 digital cross-connect systems (see Figure 5.2). In terms of maintenance and control, the existing DS3 network has numerous desirable features, such as parity bits to detect errors and signaling patterns to indicate equipment failures. Before proceeding further with details of video transmission over DS3, it is of interest to look at the hierarchy of digital transmission speeds in this network. The basic building block is a T-1 line. This is the telephone company's name for a digital line that operates at 1.544 Mbits/s. Details of this facility are discussed in the next section.

5.1.1 Digital Transmission System Hierarchy

T-1 defines a transmission rate of 1.544 Mbits/s and is generally composed of 24 or more channels of digitized voice and data. Time division multiplexing is used to interleave individual channels into the composite 1.544 Mbit/s serial stream. The DS1 format is illustrated in Figure 5.3. DS1 goes beyond the transmission rate of T-1 to define interface specifications (level, coding, pulse shape, and the like). DS1 signals are bipolar, having three states: +, 0, and −. Successive 1s (called *marks;* 0s are referred to as spaces) normally alternate in polarity (+ − + − + −). This alternation is referred to as *alternate mark inversion,* or AMI. AMI is the standard coding technique for DS1 signals.

A typical application for video transmission in a DS3 network is illustrated in Figure 5.4. The video codec (coder-decoder) is used to interface analog video and audio signals to the DS3 network. The video is mapped into a standard DS3 format by the codec. A block diagram of the codec is illustrated in Figure 5.5. There are three main stages in the codec:

- Conversion of analog signals to digital data, and vice versa
- Compression and expansion (compounding) of the digital data to a rate that will allow DS3 transmission
- Multiplexing and formatting of the digital data into a standard DS3 signal

In the following sections we will discuss details such as digitizing the video, video compression, coding, and performance standards. Before we consider these topics it is instructive to review characteristics of video signals.

5.1.2 NTSC Video

In the United States, Japan, and Canada, the term *video* commonly refers to the 525-line National Television Committee (NTSC) standard color television signal along with its associated audio signals (mono or stereo). The television signal is a composite baseband video signal with a bandwidth of 5.2 MHz.

At the transmitter, the scene is scanned by three separate pickup tubes (pixels for a CCD camera), each being sensitive to one of three primary colors, red, blue, and green. Various combinations of these colors can be mixed to form any color to

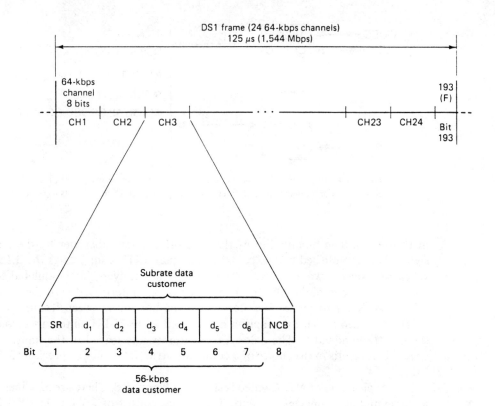

Notes:

1. Channels 1 through 24 used for data/voice channels in conjunction with the T1W84 time-division multiplexer.
2. Channels 1 through 23 used for data channels in conjunction with the T1DM time-division multiplexer. CH24 is used for fast frame sync and alarms.
3. F bit used as a DS-1 frame sync (modulo 2).
4. SR = subrate sync bit used for 2.4-, 4.8-, 9.6-kbps customer synchronization, data for 56-kbps customer.
5. Bit 1 used as a data bit for 56-kbps customers.
6. NCB = network control bit used to distinguish customer data from network control data.

Figure 5.3 DS1 transmission format. (From M. F. Slana and H. R. Lehman, "Data Communication Using the Telecommunication Network," *Computer,* May 1981. © 1981 IEEE.)

which the eye is sensitive. The three color signals are fed into transmitter signal processing circuits (matrix) where the Y, or *luminance* signal, and the *chroma,* or color signal, are created. These signals are referred to as the I and Q color difference signals. These signals are used to "quadrature amplitude"-modulate a 3.58-MHz color subcarrier. The Y signal contains just the right proportion of red, green, and blue so that it creates a normal black and white picture. The Y signal modulates the video carrier in the same manner as a signal from a normal black and white camera.

The baseband spectrum for a composite color TV transmission is illustrated

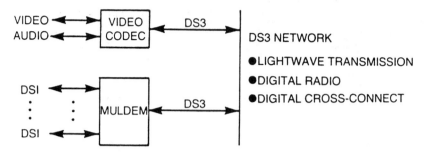

Figure 5.4 In a typical DS3 network application for video transmission, the video coder-decoder (codec) is used to interface analog video and audio signals to the DS3 network.

in Figure 5.6. The modulation of the 3.58-MHz color subcarrier by the I and Q signals is accomplished with a balanced modulator. This suppresses the 3.58-MHz subcarrier, since it would cause interference in the receiver. The modulated subcarrier sidebands containing the color information are interleaved with the Y signal's sidebands. Refer to Figure 5.7. Because of the horizontal sweep frequency (15.75 kHz), the video signal information is clustered at 15.75-kHz intervals throughout the 4-MHz bandwidth. The color information is interleaved in the unused signal space. This results in the choice of a color subcarrier frequency of precisely 3.579545 MHz.

A typical line of NTSC video is shown in Figure 5.8. There are 525 lines within a frame and 30 frames per second. The frames consist of 2:1 interlaced fields, so the field rate is 60 fields per second. When the NTSC video signal is modulated (VSB-AM) on an RF carrier for transmission, the total allocated bandwidth is 5 MHz. This includes the audio, which is modulated on a 4.5-MHz subcarried located above the video.

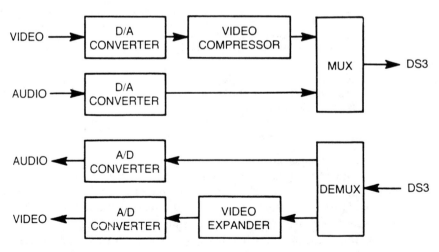

Figure 5.5 The actual video codec consists of multiplexing/demultiplexing circuits as well as A/D converters.

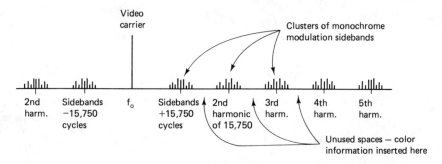

Figure 5.6 Interleaving process.

The NTSC video signal has a horizontal synchronizing pulse every 63.5 microseconds, which is the start of each line, and a series of vertical pulses every 16.7 milliseconds, which is the start of each field (Figure 5.9). The peak-to-peak voltage is 1.0 volt. A reference color burst is inserted in each line. The rest of the video signal carries the actual picture information (also called *active video*), and the signal amplitude will vary with the picture content.

The baseband audio signal has a dynamic range of 21 dB and a nominal bandwidth of 15 kHz. It is standard practice to carry up to two audio channels (stereo) with a single video. There is no additional synchronizing information associated with the audio, but to maintain lip synchronization, it is necessary to ensure that the differential delay between the video and the audio is within specified limits.

It is perhaps appropriate to point out that there is considerable interest in video signals other than the NTSC signal discussed above. For example, Europe has three principal color television standards:

1. NTSC modified for 625 lines per frame and 50 fields per second; the color difference signals are quadrature amplitude-modulated on a 3.58-MHz subcarrier with suppressed-carrier output.

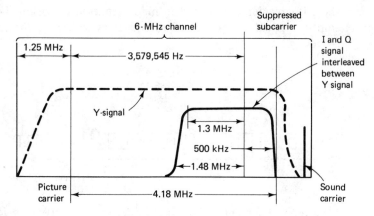

Figure 5.7 Composite color TV transmission.

Sec. 5.1   Factors Affecting Digital Video

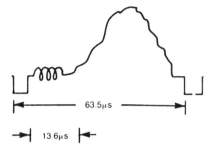

Figure 5.8 In the NTSC video signal, there is a horizontal synchronizing pulse every 63.5 microseconds. This signals the start of each video line.

2. Phase alternation line (PAL), using a 4.43-MHz subcarrier whose phase alternates from line to line.
3. Sequential with memory (SECAM), in which a 4.43-MHz subcarrier is frequency-modulated alternately by color difference signals.

PAL has a nominal video bandwidth of 5.5 MHz, whereas SECAM has a bandwidth of 6.0 MHz. Both incorporate 626 lines per frame, 50 fields per second with a 2:1 field interlace for every frame. When modulated on an RF carrier, the total bandwidth used is 8 MHz.

For studio-to-studio links, it is necessary to handle the individual color components separately to avoid the artifacts inherent in any composite signal where the luminance and chrominance are mixed. CCIR has specified a standard for this conversion, namely CCIR Rec. 601, "Encoding Parameters of Digital Television for Studios." The use of digital components will result in more flexible and higher-quality program production techniques and will facilitate the international exchange of programs. The coded components are the luminance Y and the two color difference signals R-Y and B-Y. For 525 lines and 60 fields, each line has 858 Y samples, 429 R-Y samples, and 429 B-Y samples, that is, a 4:2:2 proportion. The luminance sampling frequency is 13.5 MHz and the color difference sampling frequency is 6.75 MHz. Each sample is quantized to 8 bits with uniform PCM for a combined bit rate of 216 Mbits/s.

There is considerable interest in high definition television (HDTV), which is an attempt to provide video signals to end users with a quality matching that of

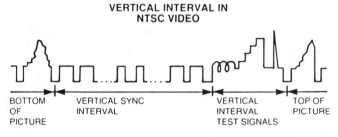

Figure 5.9 In the same NTSC video signal, there are also vertical synchronizing pulses every 16.7 microseconds at the start of each field.

photographic film. While a global standard has yet to be finalized, several candidates have emerged over the past few years. Among these are

- EIA RS-412-A
- NHK 1125-line system
- BBC 1501-line system

The field frequency is 60 Hz with an interlace ratio of 2:1, the luminance bandwidth is about 20 MHz, and the chrominance bandwidth is 6.5 MHz. With about 30 MHz of video baseband necessary per HDTV channel, it is clear that an off-the-air broadcast for a single HDTV channel would require an RF bandwidth of five to six VHF or UHF TV channels. Furthermore the HDTV format is not compatible with NTSC (that is, existing television receivers cannot reproduce a television signal for their sets from an incoming HDTV signal); hence, all existing sets would have to be replaced.

Because of this incompatibility of HDTV, other so-called compatible formats have been proposed to improve the picture clarity of existing NTSC. Two different formats for enhancing the current NTSC standard have gained some acceptance: BNTSC and MAC. BNTSC uses a composite format, similar to NTSC, with a color subcarrier. The horizontal blanking interval is replaced with digital audio signals. The luminance bandwidth is 6.0 MHz, resulting in a perceptibly sharper picture. The signal can be progressively scanned to display a 1050-line picture. MAC (multiplexed analog components) does not have a color subcarrier and instead transmits the luminance and chrominance components by compressing them to a standard video line.

Of the available 52.5 microseconds corresponding to the active video portion, 35 microseconds is used for luminance and 17.5 microseconds is used for the chrominance. Both formats are compatible in the sense that existing NTSC TV sets can reconstruct a picture with the help of a set-top box.

These formats should be viewed as interim solutions to get over the bandwidth bottleneck for HDTV. With fiber, this bottleneck does not exist. It can thus be argued that eventually three types of video formats will survive: NTSC video, because of the large number of compatible TV sets already in use, CCIR standard component TV for studio distribution, and HDTV for those who desire high-quality pictures. The data rates necessary to code and transport these signals will vary considerably, depending upon the cost and the complexity for a given bandwidth. Table 5.1 gives some representative data rates.

The following discussion deals exclusively with NTSC video and DS3 rate transmission.

5.2 ANALOG-TO-DIGITAL CONVERSION

The first step in encoding NTSC video for digital processing is the conversion of the analog signal to digital. This must be accomplished for both the baseband video and the baseband audio. The critical parameters are

TABLE 5.1 VIDEO DATA RATES

Video Type	Source Rate (Mbits/s)	Data Rate (Mbits/s)	Standard
HDTV	1000	600	SONET, OC-12
		150	SONET, OC-3
			ISDN, H4
CCIR Component	216	150	SONET, OC-3
			ISDN, H4
		45	DS3
NTSC	90	45	DS3

- Sampling frequency
- Number of bits/sample

Using Nyquist's theorem, the theoretical minimum sampling frequency for the video is 8.4 MHz (see Section 4.3), that is, twice the baseband video bandwidth. This prevents unwanted aliasing components' being introduced into the decoded video. In practice, a cutoff filter is used to band-limit the video signal prior to sampling and to filter out the aliased components after decoding.

Because of the roll-off characteristics of the filter, the practical sampling frequency is usually above the Nyquist limit, and the circuit complexity can be minimized if there is a safeguard band between the aliased components and the desired baseband spectrum. Also, in addition to alias components, unwanted components are also introduced as a result of the quantization process involved in PCM coding. This is particularly true for the composite signal. In colored areas of the picture, the visibility of this form of distortion varies as the sampling frequency is altered. The effect is minimized if the sampling frequency is an exact multiple of the color subcarrier frequency (3.58 MHz). There are two requirements, then, on the sampling frequency for the video. These are:

- It must be greater than 8.4 MHz.
- It must be an exact multiple of 3.58 MHz.

With these restrictions the choice of sampling frequency is 10.47 MHz. This is three times the color subcarrier and is locked to it.

The second parameter of interest in digitizing the video is the required number of bits per sample or equivalently the number of quantization levels. Inadequate quantization leads to the following effects:

- *Contouring.* Areas of the picture where the brightness varies slowly with position are represented by patches of uniform brightness separated by sharp transitions.
- *Beat patterns on color pictures.* Beat patterns are most noticeable in areas of constant hue and saturation. These effects can be minimized by increasing the

number of bits per sample or by locking the sampling frequency to the color subcarrier frequency.

- *Increase in noise.* Quantization noise due to an inadequate number of bits per sample. This is most noticeable in the flat parts of the picture.

The effects of quantization noise can be minimized by increasing the number of bits per sample. An example will illustrate the calculation of signal-to-quantizing noise ratio (SNR_Q).

Example 5.1

(a) Determine the theoretical signal-to-quantizing noise ratio (SNR_Q), if 8 bits/sample are used.

(b) What is the theoretical improvement per bit in the SNR_Q?

Solution

(a) $SNR_Q = \dfrac{\Delta V_s}{V_N} = \sqrt{12}(2^n - 1)$

$= \sqrt{12}(2^8 - 1) = 883.34591 \Rightarrow 58.92$ dB

(b) Assume that the number of samples n is 7.

$SNR_Q = \sqrt{12}(2^7 - 1) = 439.9491 \Rightarrow 52.86$ dB

Improvement/bit $= 58.92 - 52.86 = 6.06$ dB

The requirements for digitizing the baseband audio signal are

- A sampling frequency greater than 30 kHz. This is sufficient for an audio bandwidth of 15 kHz.
- 14 to 16 bits/sample.

For practical reasons, the sampling frequency is chosen to be higher than the required limit. The specifications for both voice and video are summarized in Table 5.2. With a 10.74-MHz video sampling rate and 8 bits/sample, the video source rate is 86 Mbits/s. For stereo audio at 32.9 kHz sampling rate and 16 bits per sample (one bit for parity), the audio source rate is approximately 3 Mbits/s. From Table

TABLE 5.2 VIDEO REQUIREMENTS

Specification	Requirement	Sample Rate	Bits/Sample
RS-250B short-haul video weighted SNR	67 dB	10.74 MHz	9
RS-250B medium-haul video weighted SNR	60 dB	10.74 MHz	8
RS-250B long-haul video weighted SNR	54 dB	10.74 MHz	7
RS-250B short-haul audio SNR	66 dB	32.9 MHz	14
RS-250B medium-haul audio SNR	65 dB	32.9 MHz	14
RS-250B long-haul audio SNR	57 dB	32.9 kHz	12

5.2, we note that the bit rate must be reduced by a factor of 2 for DS3 transmission. Since the video rate is much greater than the audio rate, only the video data is compressed prior to transmission. Techniques for accomplishing this are considered in the next section.

5.2.1 Video Compression

The basic intent of video compression is to represent the digital signal with as few bits as possible. We would suspect, of course, that as the number of bits are reduced, the information content is also reduced with a consequent degradation in picture quality. The trade-off, then, is between the complexity of compression and the resulting picture quality for a given transmission rate. There are two basic video compression schemes in use:

- Transform coding
- Differential pulse code modulation (DPCM) coding

In transform coding, a block of picture elements from one line or adjacent lines is transformed with the help of a one- or two-dimensional transform such as the Hadamard, Fourier, or discrete Fourier transform. Viewed in a mathematical setting, this represents a matrix operation where the matrix of picture elements is multiplied by the matrix of transform coefficients to obtain the matrix of transformed components. These components can be arranged in a hierarchical manner with the low-frequency (or -order) components being more important in terms of information content. We can assign more bits to the low-frequency components than are assigned to the high-frequency components. Bit rate reduction is achieved by varying the number of bits assigned to the components for transmission. As might be suspected, transform coding is complex and is generally used when a high degree of compression is required, for example, when the transmission rate is 1.5 Mbits/s or less.

The second technique, DPCM, is considerably simpler (Figure 5.10). In DPCM coding the digital samples are predicted using prior samples and the difference is quantized and transmitted to the receiver (see Figure 5.11). The receiver then reconstructs the original sample by adding the predicted value to the received digital word. Bit rate reduction is achieved by using a smaller number of bits to quantize

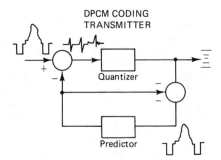

Figure 5.10 In differential pulse code modulation (DPCM), with the video quantized to 8 bits per sample, the difference signal is quantized to an average 4 bits per sample, giving the desired 2:1 compression ratio.

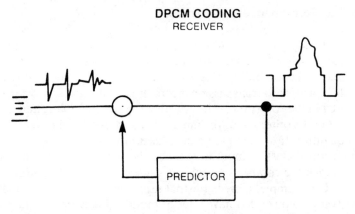

Figure 5.11 The DPCM receiver reconstructs the original sample by adding the predicted sample value to the received word.

the difference signal at the receiver. Typically, where video is quantized to 8 bits/sample, the difference signal is quantized to 4 bits/sample. This yields the desired 2:1 compression ratio. With a reasonably good prediction, the range of the difference signal is limited so that 4 bits per sample are adequate. Now in addition to this bit rate reduction technique, a nonlinear quanitizer is used. This is similar to the A-LAW and μ-LAW companding quantizers used in voice coding. This ensures that the signal-to-noise ratio is reasonably uniform for different signal levels.

While fiber optic transmission channels have a low channel error rate, it is still necessary to evaluate the effect of channel errors on picture quality. Generally, the greater the degree of compression is, the greater the susceptibility to transmission errors. Refer to Figure 5.12. To minimize errors, a channel coder for error correction and the addition of error-compensation circuitry are usually implemented with the more complex coders. The error-compensation circuitry can be as simple as replacing lost video samples by adjacent samples. An important characteristic of the decoder is that it must be able to recover quickly, since any perceptible degradation of the picture is quite noticeable.

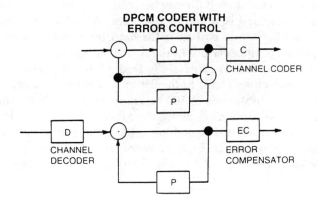

Figure 5.12 The greater the degree of compression, the greater is the susceptibility to transmission errors. In order to lessen these errors, a channel coder for error correction and some error-compensation circuitry are usually implemented with the more complex coders.

Sec. 5.2   Analog-to-Digital Conversion

5.2.2 Performance Requirements

Digital coding of video waveforms can lead to impairments that are quite unique. These include artifacts due to inadequate quantization, unlocked sampling, nonoptimized prediction and quantization in the DPCM encoder, and susceptibility of the coding process to transmission errors. It is relatively easy to correct the impairments caused in the A/D process by selecting the right sampling and quantization parameters. On the other hand, the impairments caused by video compression are difficult to quantify. If a fixed predictor is used in DPCM coding, it is generally optimized for a broad class of input waveforms. However, it is easy to find video waveforms for which the compression scheme causes noticeable impairments.

For example, a predictor that uses previous frame information does not work well with pictures that have a lot of motion. Likewise, the adverse effects of inadequate quantization in the DPCM loop can be seen in some pictures but not in others. Thus, any performance criteria for coding algorithms must include a broad range of tests to detect these impairments. As far as the susceptibility of the coding process to transmission errors is concerned, it can vary with the coding algorithm. Again, it is necessary to quantify this effect so that a standard performance requirement can be written and imposed.

5.3 DESIGNING FIBER OPTIC NETWORKS FOR HIGH SPEED

At this point, the ability of fiber to handle data rates into the gigahertz region is an established fact. This means that the fiber's I/O interfaces must be able to accommodate these data rates. As of this writing, suitable standards have not been agreed upon to facilitate this design. The Fiber Optic Distributed Data Interface (FEDDI) mentioned at the beginning of this chapter may face obsolescence before it sees general use. There is at present, however, a proposed High Speed Channel Standard (HSC) before the ANSI-X3T9·3 committee. This standard addresses point-to-point links for data rates of 800 Mbits/s and later 1600 Mbits/s. This will serve more demanding applications where FEDDI's 100 Mbit/s specification is not adequate. The HSC protocol is aimed primarily at point-to-point fiber links but also includes provisions for switched architecture networks.

In previous chapters we discussed such fiber optic areas as power budgeting, dispersion constraints, and choice of LEDs and photodiodes. A typical fast fiber link is illustrated in Figure 5.13. Note in particular the need for parallel-to-serial and serial-to-parallel conversion at the interface to the fibers. Additional detail for a typical design operating at 800 Mbits/s is illustrated in Figure 5.14. In general, the rate of flip-flop clocking is the limiting factor in the design of conversion circuits. As a general rule, TTL should be considered to 100 MHz; discrete ECL is suitable to 250 MHz, and ECL gate arrays are a good choice for conversions to 500 MHz. For applications at higher frequencies, either multiple ECL circuits in parallel or gallium–arsenide technology is most suitable. ECL is power hungry but is less expensive than gallium–arsenide base converters. These converters, however, can handle data rates in the gigabits/s region.

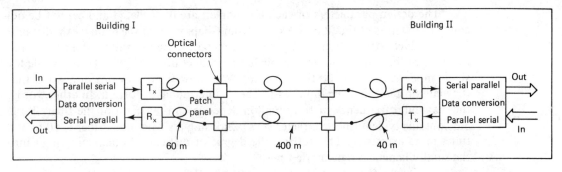

Figure 5.13 Fast fiber links of the type shown typically span less than 2 km. Most star- and ring-type networks in this configuration will be significantly simplified by the HSC protocol.

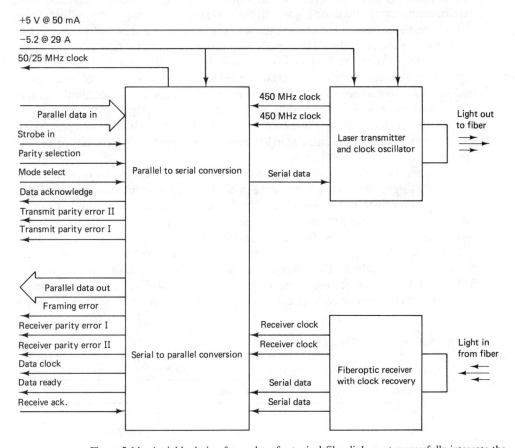

Figure 5.14 A viable design for nodes of a typical fiber link must successfully integrate the characteristics of four components: receiver, transmitter, data conversion circuitry, and cable. The data conversion circuitry is the most flexible and is the most open to customization; it must also be designed with the greatest care.

Sec. 5.3   Designing Fiber Optic Networks for High Speed

The design parameters of each component are interrelated and cannot be decoupled. In terms of the fiber, we know from Chapter 3 that the bandwidth–distance (B·L) product, attenuation, and cable type (single mode or multimode) are the parameters of interest. Multimode cable has a product up to 1200 MHz-km, while for single mode it is 50 to 100 times greater. We know from previous discussion that the main obstacle in using multimode fiber is intermodal distortion. In other words, the maximum frequency of transmission depends on the length of the fiber. Table 5.3 contains typical fiber characteristics. Choosing a single-mode fiber is 50 to 100 times more expensive. This reflects the degree of difficulty in launching light into the small-diameter single-mode fiber.

Obviously, the choice of a transmitter must be made in conjunction with selection of the fiber-cable type. Once again the distance bandwidth product plays an important role. The vast majority of links are less than one-half mile in length. In fact, the FEDDI standard specifies a maximum node-to-node distance of $1\frac{1}{4}$ miles (2 km) because this meets 90 percent of all requirements. Some applications, however, require distances of 10 km or higher. In this case the use of single-mode fiber along with single-mode lasers is dictated. Recall from Chapter 2 that both the spectral width and the wavelength of lasers vary with power output and temperature. Thus, the power level should be kept as low as possible, consistent with the power budget. A good transmitter design incorporates dynamic power level control and power feedback circuitry to prevent the transmitter from burning out. Thermionic coolers can be used when higher power outputs are necessary (see Figure 2.12). A general design rule for designers operating up to 2 Gbits/s is to apply a power penalty of about one-twentieth of the calculated attenuation in order to compensate for dispersion.

The necessary transmitter power is ultimately determined by the receiver sensitivity and its signal-to-noise performance in conjunction with the fiber's characteristics. Typical receiver architecture is a PIN-diode or avalanche photodiode detector followed by a transimpedance amplifier and band-limiting circuitry. Circuitry (bit synchronizers) following the filter recovers the original clock signals from the data. Typical figures for a system are -3 dBm for transmitter power. With a power budget of 15 to 17 dB, the receiver should operate reliably at -20 dBm for a specified BER of 10^{-9} or better. Receiver sensitivity should be viewed in a conservative manner. Although receiver sensitivity may be specified at -25 or -30 dBm, noise prob-

TABLE 5.3 FIBER CHARACTERISTICS

Parameter	Cable Type	
	Multimode	Single mode
Core size (M)	50, 62.5	9, 10
Loss (dB/km)	3–5 (50 m) 5–8 (62.5 m)	Less than or equal to 0.5
Bandwidth–distance product	700 (50 m) 250–500 (62.5 m)	To 100,000
Connector loss (dB, typ.)	0.3–0.5	0.2

lems increase. In addition to these considerations, the dynamic range of the optical signal must be considered. A short link (<200 feet) may result in an attenuation of only a few decibels, whereas a 1-km link with multiple connectors may have an additional 8- to 10-dB loss. Automatic gain control (AGC) becomes a necessity. The system must also monitor system framing along with functions such as data parity. In video applications, parity may be ignored, but framing must be maintained at all times. Optical margin testing of DS3 links is discussed in Chapter 8. The design of optical receivers is the subject of Chapter 6.

REFERENCES

5.1. Ross, F. E. 1986. FEDDI—a tutorial. *IEEE Communications Magazine* 24:10–17.
5.2. Burr, W. E. 1986. The FEDDI optical data link. *IEEE Communications Magazine* 24:8–24.
5.3. Lundgren, C. W., and P. S. Venkatesan. 1986. Applications of video on fiber cable. *IEEE Communications Magazine* 24:33–49.
5.4. Olshansky, R. 1988. Promising merger of microwave and lightwave. *The Journal of Fiber Optics*, pp. 25–30.
5.5. Killen, H. B. 1986. *Telecommunication and Data Communication System Design with Troubleshooting*. Englewood Cliffs, N.J.: Prentice Hall.

PROBLEMS

5.1. Assuming a video bandwidth of 5.5 MHz for the PAL system, determine:
   (a) The horizontal sweep rate.
   (b) The number of distinguishable pictures elements, assuming that 84 percent of each horizontal line and 90 percent of the 626 lines are visible.
   (c) Assuming that the percentages remain the same, repeat (b) for NTSC video.
5.2. It is stated in the text that a balanced modulator is used to suppress the 3.58-MHz color subcarrier. If a modulating signal $f(t) = 2\cos(2000\pi t) + \sin(4000\pi t)$ is applied to the double sideband modulator (DSB), sketch the power spectrum at the output. Refer to Figure P5.2.

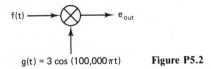

Figure P5.2

5.3. Double sideband signals can be generated by multiplying the message with a nonsinusoidal carrier and then filtering the result. Refer to Figure P5.3.
   (a) Show that this system will work if the carrier, $g(t)$, has no DC component and the cutoff frequency of the filter is $f_c + f_x$, where f_c is the fundamental frequency of $g(t)$ and f_x is the highest frequency in $f(t)$.

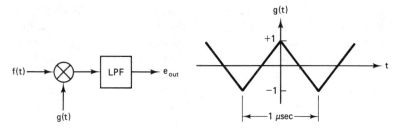

Figure P5.3

(b) For $f(t) = 3 \cos (2000\pi t)$, find the filter bandwidth. Write an expression for e_{out}.
(c) What modification would be necessary to the design if $g(t)$ has a DC component?

5.4. In Problem 5.3, assume that $g(t)$ is a rectangular waveform with period $1/f_c$ and that it is an even function with respect to t. Show that the system will demodulate a DSB signal $e_{DSB}(t) = Af(t) \cos 2\pi ft$.

5.5. Two signals $f_1(t)$ and $f_2(t)$ can be modulated onto the same carrier by using the quadrature multiplexing scheme shown in Figure P5.5.
  (a) Verify the operation of the system.
  (b) If the local oscillator at the receiver has an offset in phase of $\Delta\theta$ with respect to the transmitter, find the outputs $e_{out}(1)$ and $e_{out}(2)$. Assume $\Delta\theta \ll 1$.

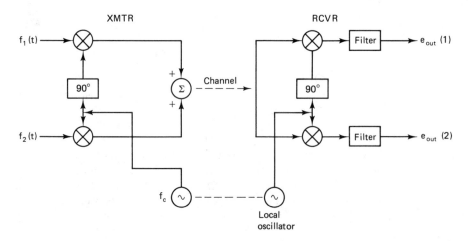

Figure P5.5

6

Optical Receivers

6.0 INTRODUCTION

Optical transmitters and receivers were discussed in Chapters 2 and 3. In this chapter we wish to consider in more detail the design of the optical receiver. We have already discussed, for example, the design of a transimpedance amplifier (see Section 2.1.2). It is of interest at this point to consider characteristics of the photodetector in greater detail. Optimization of the photodiodes' parameters is necessary in arriving at the high-speed designs needed in today's fiber optic systems.

A photodetector "demodulates" an optical signal by generating a current that is proportional to the intensity of the optical radiation, thus converting the variations in optical intensity into an electrical signal. The most important characteristics of a photodetector are efficiency, speed, noise, and physical compatibility. The efficiency of a photodetector is generally expressed in terms of "responsivity," in units of amperes/watt [see Equation (1.14)]. The responsivity represents the amount of photocurrent generated per unit of incident optical power. The speed of the photodetector must be sufficiently fast to accommodate the information rate. As noted in Chapter 2, the intrinsic noise in a photodetector is composed primarily of thermal and shot noise. Furthermore, the relative importance of shot noise and thermal noise depends on the received optical power and the value of the load resistor. As an example, in a 50-Ω system, shot noise will predominate when the averaged photocurrent exceeds about 200 μA. Suitable receiver design can be utilized so that thermal noise can be reduced, leaving shot noise to form the basic limitation in most optical communication systems. Excessive shot noise can result from high leakage and dark currents.

As already noted, the most common detectors used in fiber optic systems are PIN and avalanche photodiodes (APDs). The wavelength of interest dictates the materials used in construction of the diode. For example, silicon photodiodes can detect radiation from visible wavelengths up to about 1 μm. Germanium photodiodes respond up to about 1.6 μm. GaAs photodiodes span the range from 0.7 to 0.9 μm, and InAsP can respond up to 1.6 μm, depending on the material composition. In general, a semiconductor that responds to a longer wavelength has a lower band gap, and thus the leakage and dark currents are larger. The degree of dependence on temperature is also higher.

A PIN diode with bias is illustrated in schematic form in Figure 1.24 and again in Figure 6.1. Under reverse bias, the intrinsic region is completely depleted of carriers. A high electric field exists there, perpendicular to the layer. Initial doping of the intrinsic region determines the value of reverse bias voltage above which full depletion occurs. For example, lower doping results in a low depletion voltage.

Typical PIN photodetectors require only a few volts for full depletion of the intrinsic region; with full depletion, a photon (hf) incident on the intrinsic layer will be absorbed, and a hole-electron pair will be generated. At this point, the electric field present in this region rapidly attracts the electrons and holes to the N and P sides of the diode. If an external circuit is complete, a flow of current is produced. Also, any photons that are absorbed in the top P-layer and the N-layer will also produce free carriers. These must diffuse back into the intrinsic region before they can be collected and contribute to current flow.

The diffusion time constant can be quite long—on the order of nanoseconds. In terms of speed, this results in a serious penalty. An additional factor limiting the speed of response is the shunt capacitance associated with the depletion region. A

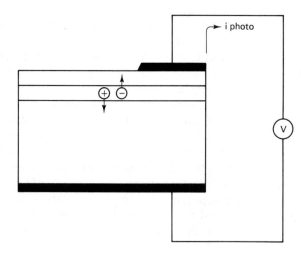

Figure 6.1 Schematic diagram of a PIN photodiode.

third (and less serious) limitation on speed is due to the finite time (drift time) required for the photogenerated carriers in the intrinsic region to be swept across the layer. This drift time is less than 5 ps in a well-designed diode, and thus does not have a significant impact for frequencies less than 100 GHz.

From the above discussion we see that in order to effect a well-designed PIN diode, we must do the following:

- Reduce doping in the intrinsic layer.
- Reduce shunt capacitance.
- Avoid optical (photon) absorption in the P and N layers.

The first step is accomplished by careful material processing. Shunt capacitance is affected by the first step. Shunt capacitance can also be reduced by reducing the size of the diode. There is a limit to this, however, where physical incompatibility becomes a problem. The third step is accomplished by using a "heterostructure" in which the top P layer is composed of GaAlAs while the intrinsic layer is made of GaAs and is sufficiently thick to absorb almost all of the incident light. The top P layer has a higher bandgap and is therefore transparent to optical radiation, which would be absorbed in the intrinsic region. The diagram for a heterostructure PIN photodiode is illustrated in Figure 6.2. Bandwidths on the order of 6 GHz have been obtained with this design. The bandwidth is obtained by first obtaining the impulse response of the photodiode (see Section 1.1.4). A Fourier transform of the impulse response then yields the frequency response of the detector.

A Schottky-barrier photodiode works on the same general principle as a PIN. Optical absorption, however, takes place at the depletion region of the metal-semiconductor Schottky junction instead of a separate I (intrinsic) layer. Due to the presence of the metal layer on the surface of the photodiode, the responsivity is somewhat lower than that of PIN diodes. Schottky-barrier photodiodes with a bandwidth on the order of 100 GHz have been produced.

The basic structure of an avalanche photodiode is the same as that of a PIN photodiode. When the reverse bias voltage to the diode is increased, the electric

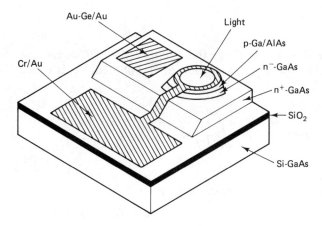

Figure 6.2 GaAs heterostructure PIN photodiode.

field in the depletion region increases corespondingly. When the voltage becomes sufficiently high ($\sim 10^5$ V/cm), an electron or a hole can collide with a bound electron with sufficient energy to cause ionization, thus creating an extra electron–hole pair. These additional carriers, in turn, gain enough energy from the electric field to cause further impact ionization. This process continues until an avalanche of carriers is produced. A single-incident photon can result in G_0 electrons and the responsivity of the avalanche photodetector is increased by a multiplication factor G_0.

The avalanche photodetector bandwidth is dependent on two factors: carrier transit time in the high field avalanche region and the rate of carrier ionization. We can reason intuitively that collision of the carriers in the avalanche process slows down the carrier speed and thus reduces the photodetector bandwidth. In fact, for high values of G_0, the relation between G_0 and bandwidth follows approximately a constant gain–bandwidth product. Device design techniques have been devised to aid in overcoming this limitation. In general, for very long-distance transmission (>100 km) where the received optical power is weak, an avalanche photodiode offers considerable advantage over PIN detectors, while the opposite is true for relatively shorter links, where the received optical power is strong.

In today's environment, further improvements in system specifications can be obtained only by a detailed understanding and improvement of devices. Optimization of parameters is a means of realizing this goal. As an example, this process is discussed in the following section for silicon photodiodes.

6.1 OPTIMIZATION OF PARAMETERS IN HIGH-SPEED SILICON PHOTODIODES

High-speed silicon photodiodes are used in optical fiber communication systems for wide-bandwidth transmission of information. The efficient employment of a high-speed silicon photodiode requires knowledge of its characteristics in order to optimize the parameters of operation. In the subsequent discussion, we will examine these parameters, one at a time, and analyze their effects on the reception of optical signals. This analysis then will be applied to practical design situations.

6.1.1 Charge-Collection Time

Refer to Figure 6.3. The time required for the electric field E to sweep out the photoexcited carriers within the depletion region of the photodiode is called the *charge-collection time*, τ_{cc}.

We can express this time as

$$\tau_{cc} = \frac{\overline{\Delta x}}{\overline{v}} \approx \frac{d}{2}\left(\frac{d}{\mu V}\right) = \frac{d^2}{2\mu V} \qquad (6.1)$$

where

d = width of the depletion region in the n-type material

$\overline{v}$ = average carrier drift velocity through the depletion region

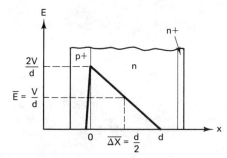

Figure 6.3 Structure of a silicon photodiode, showing the depth of the depletion region, d, into the n-type material. E is the electric field for the applied reverse bias, V, and $\overline{E}$ is the average field within the depletion region. $\overline{\Delta x}$ is the average distance of travel of the carriers across the depletion region. (Reprinted with permission from *Electro-Optical Systems Design,* December 1981, p. 38.)

μ = the carrier mobility

V = applied reverse bias

$\overline{\Delta x}$ = average distance of travel of the carriers across the depletion region

When Equation (6.1) is used, the shallow depletion region extending into the p$^+$ region is not significant. For p-on-n structures, the charge-collection time is approximately

$$\tau_{cc} \approx \frac{\rho_n}{400} \text{ ns} \tag{6.2}$$

and for n-on-p structures, it is approximately

$$\tau_{cc} \approx \frac{\rho_p}{1000} \text{ ns} \tag{6.3}$$

where ρ_n and ρ_p are the resistivities of the n-type and p-type silicon, respectively, in ohms-centimeters.

When the applied reverse bias exceeds that value for which the depletion depth just equals the thickness of the silicon chip, τ_{cc} will decrease below the values obtained from the preceding equations. For large values of the electric field E, where breakdown becomes a factor, τ_{cc} will be greater than the calculated values.

6.1.2 RC Rise Time Component

We mentioned briefly the effects of RC rise time in Section 2.5. The RC-component of rise time, τ_{RC}, is essentially the time required to discharge the photodiode junction capacitance C_J through the series combination of the external load resistance R_L and the internal series resistance of the photodiode R_S. Refer to Figure 6.4. When operating into a transimpedance operational amplifier, the effective R_L is the feedback resistance R_f divided by the open-loop gain of the op-amp (see Figure 2.8). The series resistance R_S of the photodiode is composed of three components: the resistance of the undepleted bottom region of the silicon chip, the contact resistance, and the collection resistance associated with the surface resistivity of the front diffusion. We can approximate the value of R_S from the inverse slope of the I–V characteristic in the first quadrant [Figure 6.5(a)], or from the inverse slope of the forward irradiated I–V characteristic in the fourth quadrant [Figure 6.5(b)].

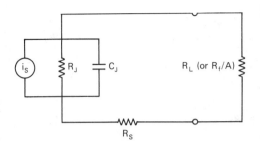

Figure 6.4 Equivalent circuit to define the RC component of rise time of the silicon photodiode. R_L is the load resistance, which, when an op-amp is used, is equal to the feedback resistance, R_f, divided by the open-loop gain of the op-amp, A. C_J is function capacitance, and I_s is the signal photocurrent in response to the incident light signal. (Reprinted with permission from *Electro-Optical Systems Design,* December 1981, p. 38.)

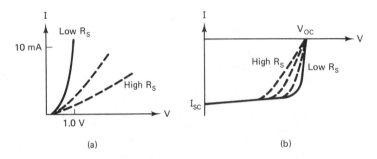

Figure 6.5 (a) Forward I-V characteristic and (b) forward irradiated I-V characteristic of the silicon photodiode. V_{OC} is the open-circuit voltage, and I_{SC} is the short circuit. (Reprinted with permission from *Electro-Optical Systems Design,* December 1981, p. 38.)

It is necessary to add to the junction capacitance C_J the MOS (metal-oxide-semiconductor) capacitance associated with any portion of the front surface metalization that extends over the passivating oxide of the planar structures C_{mos}, plus the photodiode packaging capacitance C_{pkg} and the external wiring capacitance C_{ext}. This is particularly true when C_J is small. When all these factors are taken into account, the photodiode response time between 10 percent and 90 percent of its final value can be expressed as [see Equation (2.57)]

$$\tau_{RC} = 2.2(R_S + R_L)(C_J + C_{mos} + C_{pkg} + C_{ext}) \tag{6.4}$$

For very small active areas on high-resistivity silicon-base material, typical values for R_S are in the range of a few ohms to several hundred ohms.

For p-on-n structures, the junction capacitance is described as

$$C_J = \frac{19,200A}{\rho_n^{1/2}(V_0 + V_B)^{1/2}} \, pF \tag{6.5}$$

where

A = photodiode active area, cm^2

ρ_n = resistivity in Ω-cm

V_0 = self-depletion bias (typically $\sim \frac{1}{3}$ volt)

V_B = applied reverse bias, volts

C_{mos} is approximately 7500 pF per cm² of metal over a typical thermal–oxide thickness of 5000 Å. C_{pkg} is on the order of 1 pF, while C_{ext} can range up to several picofarads.

6.1.3 Diffusion Time

Some of the carriers that are photoexcited outside the depletion region will diffuse into the depletion region before recombining. τ_{diff} is the time required for these carriers to diffuse into the depletion region. For visible and into near-ultraviolet wavelengths, most of the carriers are photoexcited near the front surface of the chip within the depletion region. Thus, τ_{diff} is not relevant. However, in the near infrared and wavelengths out to about 1.2 μm, silicon is more transparent. Carriers are photoexcited down into the bulk material. In fact, from 1.0 to 1.2 μm wavelengths, carriers are typically excited all the way to the back of the chip. Unless the silicon resistivity and applied reverse bias are sufficiently high to extend the depletion region through the entire chip thickness, the diffusion time is relevant and must be considered in the total rise time of the photodiode. The fully depleted operation is often employed at a wavelength of 1.06 μm to obtain rise times on the order of nanoseconds. This is accomplished by eliminating the relatively slow τ_{diff} component.

τ_{diff} is relevant whenever the photon absorption depth exceeds the depletion depth. For p-on-n structures, τ_{diff} may be expressed as

$$\tau_{diff}^{P/n} = \left(\frac{1}{13}\right)\left[\frac{3.0}{\alpha} - 0.54\rho_n^{1/2}(V_0 + V_B)^{1/2} \times 10^{-4}\right]^2 \tag{6.6}$$

where

τ_{diff} = diffusion time, seconds

α = silicon absorption coefficient at the wavelength of the incident light, cm^{-1}

In this expression $3.0/\alpha$ is the 95 percent absorption depth and the depletion depth d is equal to $0.54(\rho_n V_B)^{1/2}$. For n-on-p structures, the equivalent expression is

$$\tau_{diff}^{n/P} = \left(\frac{1}{36.4}\right)\left[\frac{3.0}{\alpha} + 0.32\rho_p^{1/2}(V_0 + V_B)^{1/2} \times 10^{-4}\right]^2 \tag{6.7}$$

Equations (6.6) and (6.7) apply to diffusion of carriers from the undepleted bottom portion of the silicon chip up into the depletion region. If lateral diffusion of carriers from outside the active area perimeter into the depletion region occurs, this must also be considered.

6.1.4 Total Photodiode Rise Time

We can approximate the total photodiode rise time as

$$\tau_{10}^{90} \approx (\tau_{cc}^2 + \tau_{RC}^2 + \tau_{diff}^2)^{1/2} \tag{6.8}$$

Actual values for commercially available nonavalanching PIN silicon photodiodes range from less than 1 ns, for small-area devices operating in the photoconductive (reverse bias) mode into small load resistance at wavelengths below 950 μm, up to several microseconds for large-area cells operating in the photovoltaic (zero-biased) mode. For partially depleted devices, fall times are usually longer than rise times. The rise times and fall times are essentially equal for fully depleted structures.

Avalanche silicon photodiodes (APDs) exhibit rise times down to a few tenths of a nanosecond (with gain) but are more expensive than PIN photodiodes. APDs, however, require careful control of operating bias and temperature (see Chapter 2). Nonavalanching PIN silicon photodiodes with rise times less than 0.1 ns are theoretically possible and are in the product development stage.

6.1.5 Amplifier Design Considerations

Refer to Figure 6.6. The output signal voltage v_s is given by

$$v_s = -i_s Z_f = -P_{inc} R_\lambda Z_f \text{ V} \tag{6.9}$$

where

P_{inc} = incident optical power

R_λ = responsivity, A/W

Z_f = feedback impedance across the op-amp

For sufficiently low frequency, this expression reduces to

$$v_s = P_{inc} R R_f \tag{6.10}$$

The upper -3-dB frequency is usually determined by the feedback loop of the op-amp and is given by

$$f_{-3 \text{ dB}} = \frac{1}{2\pi R_f C_f} \tag{6.11}$$

The associated 10–90 percent rise time τ_{10}^{90} then has the form

$$\tau_{10}^{90} = 0.35/f_{-3 \text{ dB}} \tag{6.12}$$

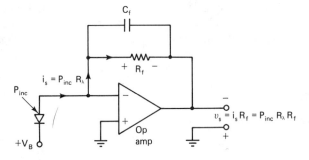

Figure 6.6 Photodiode circuit with an op-amp. The optical power P_{inc} incident on the diode surface having responsivity R generates a signal current I_s. The op-amp, with feedback resistance R_f and feedback capacitance C_f, causes an output voltage signal v_s.

Using Equations (6.10) through (6.12), we can express the incident power as

$$P_{\text{inc}} = \frac{v_s}{R_\lambda R_f} = 2\pi B C_f v_s / R_\lambda \qquad (6.13)$$

where B is the operating bandwidth in hertz.

Noise equivalent power, NEP, is the minimum detectable incident light power for which the signal-to-noise ratio is unity. That is, $v_s = v_n$, where v_n is the rms value of the noise. Thus, for unity signal-to-noise

$$P_{\text{inc}} = \text{NEP} = 2\pi B C_f V_n / R_\lambda \qquad (6.14)$$

for the wide bandwidth photodiode op-amp combination. In order to achieve a wide-bandwidth, high-speed system, the value of C_f is maintained as small as possible consistent with stable nonoscillating operation of the operational amplifier. This allows R_f to be as large as necessary consistent with the required bandwidth. An op-amp should be selected with a fast slew rate for a large gain–bandwidth product (GBP). The following examples illustrate typical design situations.

Example 6.1

A standard silicon photodiode operates at a wavelength of 632.8 nm into a load of 50 ohms. The diode is biased at 50 volts. The following parameters apply:

$$\text{Active area } A = 0.01 \text{ cm}^2$$

$$\text{Total area } A_T = 0.02 \text{ cm}^2$$

$$\text{Thickness } t = 0.040 \text{ cm}$$

$$\rho_n = 10 \text{ } \Omega\text{-cm}$$

$$C_{\text{pkg}} \approx 1.0 \text{ pF}$$

$$V_0 \approx \text{neglible}$$

$$\text{Absorption coefficient } \alpha = 4 \times 10^3 \text{ cm}^{-1}$$

Calculate the total 10 to 90 percent photodiode rise time. Assume that C_{pkg} and C_{mos} may be ignored.

Solution The charge collection time is

$$\tau_{\text{cc}} = \frac{\rho_n}{400} = \frac{10}{400} = 0.025 \text{ ns}$$

The junction capacitance is

$$C_J = \frac{19{,}200 \, A}{\rho_n^{1/2}(V_0 + V_B)^{1/2}} = \frac{(19{,}200)(0.01)}{(10 \times 50)^{1/2}} = 8.6 \text{ pF}$$

The depletion depth d is given by

$$d = 0.54(\rho_n V_B)^{1/2} = 0.54(10 \times 50)^{1/2} = 12 \text{ } \mu\text{m}$$

The series resistance R_S from the undepleted bottom region of the chip is given by

$$R_S = \rho_n(t - d)/A'$$

where A' is approximately the total chip area. Since d is much less than t, this results in

$$R_S = (10)(0.04)/0.02 = 20$$

Thus
$$C = C_J + C_{pkg} = 8.6 + 1 = 9.6 \text{ nF}$$

$$\tau_{RC} = 2.2 (R_S + R_L)C$$
$$= 2.2(20 + 50)\, 9.6 \times 10^{-9} = 1.48 \text{ ns}$$

The 95 percent absorption depth $= 3.0/\alpha = 3.0/4 \times 10^3 = 7.5\ \mu\text{m}$. Since this is less than the depletion depth d, $\tau_{diff} = 0$. Thus, the 10 to 90 percent rise time is

$$\tau_{10}^{90} = (\tau_{cc}^2 + \tau_{RC}^2 + \tau_{diff}^2)^{1/2}$$
$$= (0.025^2 + 1.48^2 + 0^2)^{1/2} = 1.5 \text{ ns}$$

Example 6.2

Repeat Example 6.1 with the following parameters:

$$\lambda = 904 \text{ nm}$$
$$V_B = 50$$
$$R_L = 50$$
$$\text{Active area} = 0.056 \text{ cm}^2$$
$$\text{Total area} = 0.11 \text{ cm}^2$$
$$t = 0.040 \text{ cm}$$

Solution

$$\tau_{cc} = 400/400 = 1 \text{ ns}$$
$$C_T = (19{,}200)(0.056)(400 \times 50)^{1/2} = 7.7 \text{ pF}$$
$$C = C_J + C_{pkg} = 7.7 + 1 = 8.7 \text{ pF}$$
$$d = 0.54\,(400 \times 50)^{1/2} = 76.4\ \mu\text{m}$$
$$R_S = (400)(0.32/0.11) = 116\ \Omega$$
$$\tau_{RC} = 2.2(116 + 50)(8.7 \times 10^{-12}) = 3.18 \text{ ns}$$

The absorption depth at 904 nm is 75 μm. This is less than the depletion depth, 76.4 μm. Thus,

$$\tau_{diff} = 0$$
$$\tau_{90}^{10} = (1^2 + 3.18^2)^{1/2} = 3.3 \text{ ns}$$

Example 6.3

Calculate the effect on the rise time in Example 6.2 if the reverse bias is reduced from 50 V to 5 V.

Solution

$$\tau_{cc} = 400/400 = 1 \text{ ns}$$
$$C_T = (19{,}200)(0.056/400 \times 5)^{1/2} = 24 \text{ pF}$$
$$C = C_J + C_{pkg} = 24 + 1 = 25 \text{ pF}$$
$$d = 0.54(400 \times 5)^{1/2} = 24 \text{ μm}$$
$$R_S = (400(0.038/0.11) = 138$$
$$RC = 2.2(138 + 50)(25) = 10.3 \text{ ns}$$

Since the absorption depth, 75 μm, is greater than the depletion depth, 24 μm, it is necessary to include τ_{diff}.

$$\tau_{diff} = (1/13)[3.0/\alpha - 0.54(\rho_n V_B)^{1/2} \times 10^{-4}]^2$$
$$= (1/13)[(75 - 24) \times 10^{-4}]^2 = 2000 \text{ ns}$$
$$\tau_{10}^{90} = (1^2 + 10.3^2 + 2000^2)^2 = 2 \text{ μs}$$

The results of Example 6.3 illustrate the importance of applying sufficient reverse bias to obtain short rise times. Short rise times are achieved by ensuring that the absorption depth does not exceed the depletion depth.

Example 6.4

Calculate the NEP for the circuit of Figure 6.6 if the following parameters apply:

$$R_\lambda = 0.5 \text{ A/W}$$
$$C_f = 3 \text{ pF}$$
$$B = 35 \text{ MHz}$$
$$V_n = 2 \text{ mV (rms) at the op-amp output}$$
$$R_f = 1470 \text{ Ω}$$

Solution Using Equation (6.14), we obtain

$$\text{NEP} = \left(\frac{1}{R_\lambda}\right) 2\pi B C_f V_n$$
$$= \left(\frac{1}{0.5}\right)(2\pi)(35 \times 10^6)(3 \times 10^{-12})(2 \times 10^{-3}) = 2.6 \text{ μW}$$

6.2 OPTICAL RECEIVERS

In the preceding chapters we discussed the theoretical performance limitations of the optical source and fiber, which have a direct impact on the required received optical power for a given BER or SNR. In designing fiber optic links it is standard practice to specify receiver sensitivity as the average optical power input to achieve a BER = 10^{-9}, although other BERs may be used in actual practice. In order to assess the impairments of the foregoing effects, it is necessary to describe the inherent sensitivity of typical optical receivers.

There are currently four common types of receiver designs available. These are:

1. PIN photodiode and integrating front-end
2. PIN photodiode and transimpedance amplifier
3. Avalanche photodiode and transimpedance amplifier
4. Avalanche photodiode and integrating front-end

Variations on these receiver designs are illustrated in Figure 2.8. The choice of the first-stage device type (amplifier) is dependent on the application. It is usually a field effect transistor; however, bipolar devices are more convenient for lower-sensitivity applications. Generally speaking, the optical detector is a silicon device for 850-nm and 900-nm applications, while InGaAs PIN diodes and Ge avalanche devices find application at 1300 nm and 1500 nm.

The high-impedance design (Figure 2.8) is often referred to as the *integrating amplifier*. This is due to the RC time constant. The integrating front-end system consists of a photodetector supplying an input current to a high-input-impedance circuit. Because of the long RC time constant, the signal is integrated. Thus, the signal must be subsequently equalized (differentiated) to regain its waveshape and prevent intersymbol interference in a digital system. This circuit is further illustrated in Figure 6.7. Because of the high input impedance, the amplifier's input thermal noise spectral density is low [see Equation (2.8)]. Thus, the integrating front-end amplifier provides the greatest sensitivity of the three types of postamplifiers illustrated.

The transimpedance design is very popular. In this case, the photodetector supplies current into a high open-loop amplifier with feedback. As the name implies, the signal current, i_{in}, entering the amplifier's input port is converted to an output voltage V_o. In a well-designed circuit, this transimpedance (V_o/i_{in}) is simply the value of the feedback resistor R_f. Consequently, the noise current density, if the rest of the system were noiseless, is simply that generated by the feedback resistance. Furthermore, the bandwidth (again in an otherwise perfect system) is set by the amplifier input RC time constant, where R is now the feedback resistance R_f divided by the open-loop gain, Av, of the op-amp. This circuit provides a good compromise between receiver sensitivity and dynamic range. The sensitivity is 3 to 10 dB less, but there is a much greater dynamic range and no inherent baseline wanders. However, at very high frequencies (≥ 300 MHz), stray capacitance associated with the feedback resistance must be compensated for. At these frequencies, the complications in design approach those of the high-impedance design.

For very wide bandwidths or at very high frequencies (UHF and microwave) a simple resistive input (usually 50 ohms) design is used. The detector load in this case is the 50-ohm input in parallel with whatever circuit capacitance may be needed. Often this is just the stray circuit capacitance. The input noise current is determined primarily by the thermal noise of the 50-ohm resistor and the amplifier noise figure. Even though this design has the poorest noise figure, it exhibits very good dynamic range.

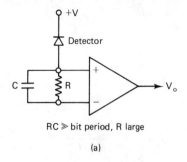

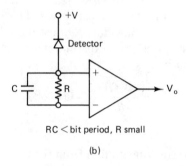

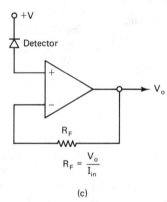

Figure 6.7 Three types of receiver circuits. (a) Integrating, (b) RC input; (c) transimpedance. (From H. R. Rice and G. E. Keiser, "Application of Fiber Optics to Tactical Communication Systems," *IEEE Communications Magazine,* May, 1985. © 1985 IEEE.)

6.2.1 Noise Considerations

The performance of the receiver preamplifier, to a large degree, determines nearly the entire receiver performance. The main characteristics to be considered are bandwidth, noise, and dynamic range. We know from the previous discussion that the choice of amplifier topology involves inevitable compromises. The manner in which receiver input impedance and photodiode capacitance affect receiver performance is illustrated in Figure 6.8. If we assume no other limiting factors inside the amplifier, the maximum achievable receiver bandwidth is

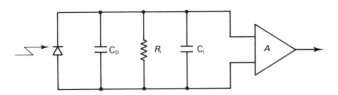

Figure 6.8 Preamplifier noise model.

$$f_{-3\,\text{dB}} \geq \frac{1}{2\pi R_i(C_i + C_D)} \qquad (6.15)$$

where R_i is the preamplifier input resistance. This also includes any other resistive load on the photodiode. C_i is the input capacitance and C_D is the photodiode capacitance. In order to increase the bandwidth, R_i must be reduced for a total capacitance. However, the penalty is increased thermal noise. From Equation (2.10), the input noise current is given by

$$\overline{i_{\text{tot}}^2} = \frac{4kTB}{R_i} + 2eB(I_p + I_d) \qquad (6.16)$$

where

$\overline{i_{\text{tot}}^2}$ = total noise current contribution from the photodiode and the preamplifier input stage

k = Boltzmann's constant

B = bandwidth, Hz

e = electron charge, coulombs

I_p = photocurrent

I_d = dark current

Note that the lower the value of R_i is, the greater the thermal noise contribution.

A FET preamplifier can be used to keep R_i very large. As noted previously, this reduces the response time and consequently the bandwidth. A further bandwidth reduction may be due to the use of a FET when it is used as a transconductance (current output) amplifier loaded with a capacitor. Under these conditions, it is necessary to use an equalizer network. This type of amplifier has high gain at low frequencies. Consequently, it can be easily overloaded and its dynamic range is severely limited.

The transimpedance amplifier (Figure 6.7) has a low output impedance, which helps to achieve wide dynamic range. Its frequency response is greater than that of a bipolar or FET amplifier. The -3-dB bandwidth may be shown to be

$$f_{-3\,\text{dB}} \geq \frac{A}{2\pi R_f(C_i + C_D)} \qquad (6.17)$$

Because the feedback resistor R_f also contributes to input stage noise, the noise performance of transimpedance amplifiers is somewhat less than FET or bipolar amplifiers that have been optimized for noise performance.

Example 6.5

The following parameters apply for the circuit of Figure 6.6:

$R_f = 5\text{M}\Omega$ $\qquad I_p = 2\ \mu\text{A}$

$C_f = 0.5\ \text{pF}$ $\qquad I_d = 1.6 \times 10^{-12}$

$C_i = 1.5\ \text{pF}$ $\qquad A_{ol}(\text{op-amp}) = 90\ \text{dB}\ (13{,}622.77)$

$C_D = 1.0\ \text{pF}$ $\qquad R_L = 0.5\ \text{A/W}$

$R_i = 500\ \text{k}\Omega$

Assume room temperature.

(a) Draw the amplifier noise model.

(b) Calculate the -3-dB bandwidth.

(c) Calculate the total input noise current.

(d) Calculate noise equivalent power.

Solution

(a)

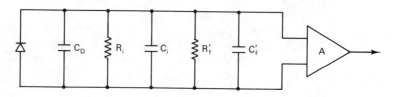

Figure 6.9

$$V_0 = i_f Z_f \rightarrow i_f = \frac{V_0}{Z_f} = \frac{A v_i}{Z_f}$$

$$Z_i = \frac{v_i}{i_f} = \frac{Z_f}{A} = \frac{1}{A}\left[\frac{(R_f)(1/j\omega C_f)}{R_f + 1/j\omega C_f}\right]$$

$$= \frac{R_f}{A}\left[\frac{1}{1 + j\omega_f C_f R_f}\right] \Rightarrow R'_f = \frac{R_f}{A};\qquad C'_f = A C_f$$

(b)

$$f_{-3\text{dB}} = \frac{A}{2\pi R_f \| R_i (C_i + C_D + A C_f)}$$

$$= \frac{31{,}622.77}{(2\pi)(5 \times 10^6 \| 5 \times 10^4)[1.5 + 1 + 31{,}622.77(0.5)]10^{-12}}$$

$$= 700\ \text{kHz}$$

(c)
$$\overline{i_{tot2}} = \frac{4kTB}{R_i} + 2qB(I_p + I_d)$$

$$= \frac{(4)(1.38 \times 10^{-23})(300)(7 \times 10^5)}{5 \times 10^5}$$
$$+ 2(1.6 \times 10^{-19})(7 \times 10^5)(2 \times 10^{-6} + 1.6 \times 10^{-12})$$

$$= 47.11 \times 10^{-20}$$

$$\overline{i_{tot}^2} = 6.8 \times 10^{-10} \text{ A}$$

(d)
$$\text{NEP} = \frac{v_0}{A} = \frac{i_{tot} \cdot R_f}{R_f \cdot R} = \frac{i_{tot}}{R} = \frac{6.86 \times 10^{-6} \text{ A}}{0.5 \text{ A/W}} = 13 \text{ }\mu\text{W}$$

6.2.2 Critical Areas of Receiver Design

With reference to the receiver designs discussed in the previous section, the two primary designs in widespread use are the high-impedance or integrating configuration and the transimpedance technique. For operation in the 1.3-μm to 1.5-μm region, the detectors may be PIN photodiodes or germanium avalanche devices. A typical material for PIN diode fabrication consists of either GaInAs/GaAs for use at 1.3 μm or preferably InGaAsP/InP for high speeds. The latter has low capacitance and high quantum efficiency over the wavelength range from 0.95 μm to 1.65 μm.

Germanium avalanche photodiodes may also be fabricated, but these diodes suffer from a number of drawbacks. The main practical drawbacks for sensitive receiver applications are

- Most commercially available devices have a large dark current at room temperature.
- The avalanche multiplication process in germanium has a large excess noise factor of 1 (compared to 0.5 for silicon). This limits the maximum usable avalanche gain.
- The required avalanche bias voltage is between 30 to 40 volts. This may be inconvenient for some receiver applications.
- The absorption depth for Ge is approximately 1.8 μm at room temperature and has a temperature coefficient of typically 3×10^4 eV/°C. At 1.3 μm, this effect is neglible. At a transmission wavelength of 1.5 μm, the quantum efficiency may exhibit some sensitivity.
- In Ge avalanche photodiodes of 100 μm and above, the distribution of gain is often nonuniform.

Decision thresholds for optical receivers with avalanche gain are dependent on the actual avalanche gain and the optical signal level. If stable receiver performance is to be achieved when avalanche gain is present, some form of bias control is necessary. This may involve a servo system that monitors signal-independent excess noise and maintains the avalanche gain at an optimum level. For the highest sensitivities, some form of temperature stabilization may be necessary (see Figure 2.12).

Nonavalanche photodetection is comparatively free from these drawbacks. The primary requirement is a high quantum efficiency at the transmission wavelength. In addition, the capacitance must be kept below 0.5 pF for good sensitivity.

Subsequent to the basic optical detection process, the most critical design aspects prior to the decision point relate to the provision of correct equalization and stable voltage gain. The actual degree of equalization needed depends strongly on the type of receiver used. For the high-impedance design receiver, exact compensation for the integration present at the input must be provided. For this, a simple differentiation network or a single-tap transversal equalizer may be used. In either case, the choice of network parameters is determined by the system bit rate and thus is subject to some degree of optimization. Transimpedance designs are inherently wideband and, in principle, require only equalization for optimum pulse shape prior to the decision point. For practical systems, the familiar raised-cosine frequency response characteristic may be chosen (see Section 6.2.3). Ideally, in a careful design, the overall system frequency response should be considered. Thus, nonideal optical pulse-shaping and inadequate system bandwidth can, theoretically, be compensated for by the overall equalization process. Further refinements may include decision-feedback equalization.

6.2.3 Optical Receiver Design Based on Nyquist's First and Second Criteria

Receivers for direct-detection optical communication that satisfy the Nyquist criteria generally employ a signal element waveform that provides a close approximation to a full raised cosine Fourier spectrum. For conventional digital electrical communications, this signal satisfies Nyquist's first and second criteria for distortionless transmission. This provides both zero intersymbol (ISI) and zero telegraph distortion (TD), so that the system eye pattern is wide open both vertically and horizontally. In this analysis, Nyquist postulated several known distortions of an incident rectangular signal that would still allow it to be recognized at the receiver. He then evaluated the bandwidth and band shaping required. Nyquist actually evaluated the transmission requirements for three criteria. The one of interest to us is illustrated in Figure 6.10. With this criterion, the transmission path is arranged for correct reproduction of the incident signal amplitude at the exact center (sampling time of each element, t_1, t_2, t_3, and so on. Note in Figure 6.9(b) that the transmission path has constant amplitude, a, up to a sharp cutoff at $f_c = 1/2T_1$. The phase shift b is linear over the same frequency range. That is, Nyquist proved that to transmit N signal elements per second, a bandwidth of $N/2$ Hz was theoretically sufficient.

For optical communications based on direct detection receivers employing avalanche photodiodes (APDs), the optimum decision threshold may be depressed significantly below the mid-range of the vertical eye opening (see Figure 3.19). In those circumstances, a full raised-cosine signal does not actually provide zero TD, since

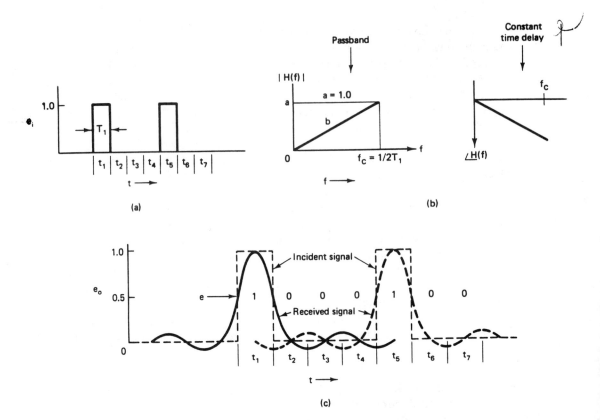

Figure 6.10 Nyquist criteria.

signal transitions, while crossing the half-amplitude level midway between sampling instants, give rise to "threshold" crossings that are displaced from these times. This results in two important consequences for practical applications:

- For untimed or 2R repeaters, providing signal reshaping and regeneration but not retiming, a nonoptimum midway decision threshold must be adopted if severe telegraph distortion is not to occur.
- For retimed or 3R repeaters (reshaping, regeneration, and retiming) the eye closes asymmetrically for ones and zeros in the presence of sampling time offset or sampling alignment jitter.

6.2.3.1 Signal Design for APD Receivers. A simplified design of an optical repeater is shown in Figure 6.11. The dashed line corresponds to the timing path for a 3R repeater. It is shown in texts on digital communications that for zero ISI, the Fourier spectrum of the equalized pulse at the input of the decision circuit must be given by

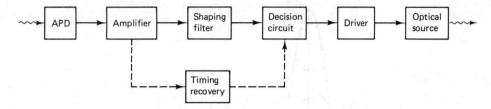

Figure 6.11 Block diagram of an optical repeater. Broken line shows the timing signal path for a 3R design.

$$H_0(f) = \begin{cases} 1 + 2d \cos(\pi f T), & |f| < \frac{1}{T} \\ 0, & \text{elsewhere} \end{cases} \quad (6.18)$$

where T is the pulse width.

The corresponding time-domain signal element waveform is described by

$$h_0(t) = \frac{\sin(2\pi t/T)}{2\pi t/T} \left(\frac{1 - 4(t/T)^2(1 - 2d)}{1 - 4(t/T)^2} \right) \quad (6.19)$$

For $d = 0.5$, the full raised-cosine response is obtained. This latter case is appropriate for optical receivers using PIN diodes in which signal-dependent noise is negligible. That is, a threshold value corresponding to one-half of the vertical eye aperture is appropriate. For APD receivers, the optimum value of d is generally less than 0.5, often significantly so. The time and frequency domain representations for $d = 0.3$ are illustrated in Figure 6.12.

The responses described by Equations (6.18) and (6.19) are the ideal minimum bandwidth solution. There are, however, practical difficulties that make these responses inappropriate for signal waveform design. Specifically, for $d < 0.5$, the spectrum falls abruptly to zero at $f = 1/T$. This implies a very sharp cut-off filter, which is difficult to realize physically. It is common practice to modify this to provide a gradual roll-off response. Mathematically, we can characterize the frequency response, $A(f)$, of the modified roll-off filter as

$$A(f) = \begin{cases} 1, & 0 < f < f_1 - f_x \\ \frac{1}{2}\left[1 - 2d \sin \frac{\pi}{2\alpha}\left(\frac{f}{f_1} - 1\right)\right], & f_1 - f_x < f < f_1 + f_x \\ 0, & f > f_1 + f_x \end{cases} \quad (6.20)$$

where

$$\alpha = \frac{f_x}{f_1} = \text{roll-off factor} \quad (6.21)$$

Here α is the ratio of the excess bandwidth of the composite channel to the Nyquist bandwidth, $1/2T$. The phase shift of the filter is given by

$$\phi(f) = Kf \quad (6.22)$$

Sec. 6.2 Optical Receivers

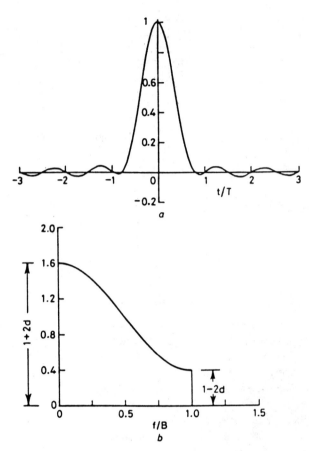

Figure 6.12 Signal designs achieving zero ISI and zero TD for $d - 0.3$ in the minimum bandwidth of $1/T$. (a) Time domain response. (b) Frequency response.

where K is constant. A voltage impulse exhibits a flat infinite bandwidth spectrum and constant phase. Thus the impulse response of a filter is the amplitude and phase response of the filter. This output in the time domain, when the filtering is a raised cosine ($d = 0.5$) is a pulse whose shape depends on the roll-off factor [see Figure 6.13(a)] between the peaks of the received signals. The corresponding frequency response of the raised cosine filter is shown in Figure 6.13(b).

An impulse is not necessary to create the responses shown in Figure 6.13. Any pulse that has a spectrum whose amplitude exceeds zero between DC and frequency $f_1 + f_x$ and whose phase remains constant over the same frequency interval can be filtered to yield those same responses. Thus, this pulse can also be transmitted at rate $2f_1$, without ISI. For example, the full-length rectangular pulses shown in Figure 6.14 fulfill these requirements. We see then the application where the input signal is data rather than an impulse. Our problem then reduces to designing a shaping filter that approximates the modified filter response. Various networks are in use for this function (see Example 6.6).

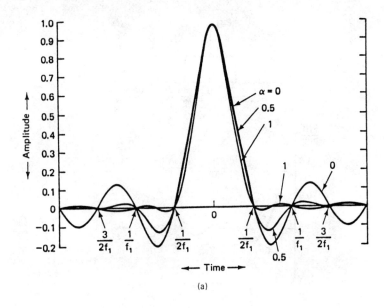

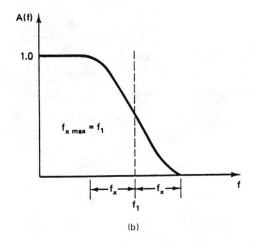

Figure 6.13 (a) Impulse response obtained in filtering is raised cosine for various roll-offs. (b) Raised cosine filter frequency response.

Example 6.6

A fourth-order Butterworth filter is shown in Figure 6.15.
(a) Find the transfer function and determine the -3-dB cutoff frequency.
(b) Plot the magnitude of the transfer function (Bode plot) and compare this with a raised cosine filter ($\alpha = 0.5$). How closely does this filter approximate the theoretical curve?

Sec. 6.2   Optical Receivers   181

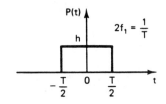

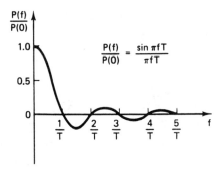

Figure 6.14 Spectral amplitude function of rectangular full-length pulse.

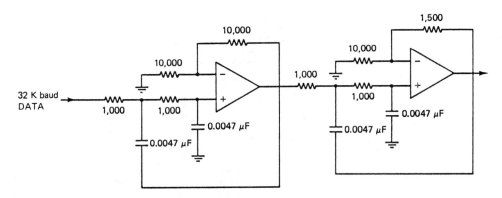

Figure 6.15

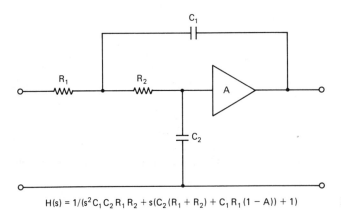

$H(s) = 1/(s^2 C_1 C_2 R_1 R_2 + s(C_2(R_1 + R_2) + C_1 R_1 (1 - A)) + 1)$

Figure 6.16

Solution

(a) The filter consists of two cascaded 2-pole low-pass Sallen-Key filters. See Figure 6.16.

For the first section: For the second section:

$A_1 = 10{,}000/10{,}000 = 1.0$ $A_2 = 1500/10{,}000 = 0.15$

$$H_1(s) = 1/(22.1 \times 10^{-12}s^2 + 9.4 \times 10^{-6}s + 1)$$

$$H_2(s) = 1/(22.1 \times 10^{-12}s^2 + 13.3 \times 10^{-6}s + 1)$$

$$H(s) = H_1(s) \cdot H_2(s)$$

(b) The plot is shown in Figure 6.17.

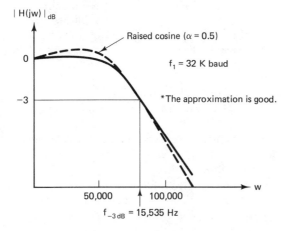

Figure 6.17

We noted earlier the importance of receiver design. It is, in fact, critical to system operation. In a digital system, there are three main performance parameters:

- Error performance
- Jitter performance
- Availability of the system

Regeneration of the signal is carried out every 35 to 40 km in a repeatered system. In practice, the sampling instants will deviate from the ideal center, and jitter is introduced (see Figure 6.10). This results in jitter, along with a consequent higher error probability. Once started, jitter propagates along the chain of repeaters. The system availability is defined to be the time that the system error probability is equal to or less than the design value. Generally speaking, the error rate should not exceed 10^{-9}. An error rate of 10^{-12}, however, is often achievable in a fiber optic system.

In this chapter, we have considered the design of receivers for direct detection systems. Coherent detection is also an option. Although presently in the laboratory stage, coherent systems offer exciting possibilities in future systems. We mentioned briefly this technique in Chapter 2. A detailed discussion is presented in Chapter 7.

REFERENCES

6.1. O'REILLY, J. J., and R. S. FYATH. 1988. New APD-based receivers providing tolerance to alignment jitter for binary optical communications. *IEEE Proceedings* 135:119–125.

6.2. KILLEN, H. B. 1988. *Digital Communications with Fiber Optics and Satellite Applications.* Englewood Cliffs, N.J.: Prentice Hall.

6.3. DUNCAN, F. 1981. Optimization of parameters in high-speed silicon photodiodes. *Electro-Optical Systems Design,* 38–43.

PROBLEMS

6.1. In the preceding discussions of optical photodetectors, several definitions are used. "Responsivity" is the ratio of detected output current to incident power. "Noise equivalent power" (NEP) is the amount of incident power per unit bandwidth needed to produce an output power equal to the detector output noise power. "Detectivity" (D) is defined as 1/NEP. The parameter D^* (called D-star) is the detectivity of a one-square centimeter detector.
(a) Derive the relation between NEP and responsivity.
(b) Show that $D^* = (A_r B_o)^{1/2}/\text{NEP}$, where B_o is the optical bandwidth and A_r is the receiver area.

6.2. NRZ data is to be transmitted through a raised cosine filter. The data rate is 32 kbits/s. We wish a minimum of 30 dB attenuation at a frequency of 1.5 times the Nyquist bandwidth. Calculate the value of α.

6.3. (a) Discuss the four main amplifier configurations currently used for optical communications systems. Comment on their relative merits and drawbacks.
(b) A high-impedance integrating front-end amplifier is used in an optical fiber receiver in parallel with a detector bias resistor of 10 MΩ. The effective input resistance of the amplifier is 6 MΩ, and the total capacitance of the amplifier and the detector is 2 pF. It is found that the detector bias resistor may be omitted when a transimpedance front-end amplifier design is used with a 270-kΩ feedback resistor and an open-loop gain of 100. Compare the bandwidth and thermal noise implications of the two cases. Assume an operating temperature of 290°K.

6.4. Discuss briefly the sources of noise in optical receivers. Describe in detail what is meant by quantum noise. Discuss this phenomenon with regard to:
(a) digital transmission
(b) analog transmission
Give any relevant mathematical formulae.

6.5. For a photovoltaic detector (bias voltage = 0), the voltage is

$$V = (kT/q) \ln\left(1 + \frac{i_{sc} + i}{i_s}\right)$$

where

i_s = reverse saturation current

i_{sc} = short circuit current

Plot the i-V curve for i_{sc} = 100 mA, i_s = 1.0 nA, i and $-i_{sc}$ = 0.

6.6. Assume that the detector of Problem 6.5 has an internal resistance of 1 Ω so that the voltage V is reduced by the iR drop. Repeat the i–V curve for this situation and compare it with the detector in Problem 6.5.

6.7. Equalization in an optical receiver may be accomplished with the circuit shown in Fig. P6.7. The equalized frequency response is shown. In this design $R_{in} = 5$ kΩ and the combined capacitance of the photodiode and amplifier is 5 pF. An upper 3-dB corner frequency for the amplifier-equalizer combination of 30 MHz is desired.
(a) Obtain expressions for ω_1 and ω_2.
(b) Assuming that $R_2 = 100$ Ω and $R_o = 5$ Ω, determine values for R_1 and C_1 such that the 3-dB corner frequency is 30 MHz.
(c) What is the corner frequency without the equalizer?

6.8. The voltage amplifier of a fiber optic system is to be designed with its input impedance matched to a detector bias resistor of 200 ohms. Calculate the following:
(a) The maximum bandwidth that may be obtained without equalization if the total capacitance is 10 pF (see Problem 6.7).
(b) Using the bandwidth from (a), calculate the thermal noise existing in this configuration. Assume a temperature of 290°K and that the thermal noise produced by the voltage amplifier results from the input impedance of the device.
(c) Replace the voltage amplifier by a transimpedance amplifier with a 10 kΩ feedback resistor and an open-loop gain of 75. Assume that the feedback resistor is used to bias the detector and that the total capacitance remains the same. Repeat (a) and (b) and compare the results.

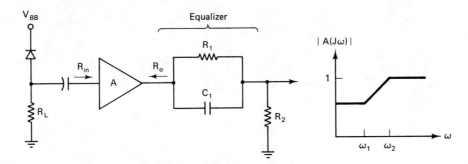

Figure P6.7

6.9. The effective input resistance of the op-amp in Problem 6.7 is 6 MΩ. Total capacitance for the amplifier and the detector is 2 pF. Calculate the following:
(a) $f_{-3\,dB}$ bandwidth
(b) Assume an operating temperature of 290°K and calculate the thermal noise referred to the input (neglect the equalizer network).

6.10. A PIN photodiode operates with the following parameters:
- Wavelength = 0.83 μm
- Quantum efficiency = 50%
- Dark current = 0.5 nA
- Temperature = 290°K

The device is unbiased but is loaded with a current mode amplifier with a 50-kΩ feedback resistor and an open-loop gain of 50. The capacitance of the photodiode is 1 pF, and the input capacitance of the amplifier is 6 pF. Calculate the incident optical power required to maintain an SNR of 50 dB, given that the postdetection bandwidth is 10 MHz. Is equalization necessary?

7

Coherent Optical Communications

7.0 INTRODUCTION

The optical communication systems discussed so far have used simple intensity-modulated transmitters along with receivers that are sensitive to optical power. The competitive performance of these optical systems has been obtained with detectors that respond to optical power rather than the optical electromagnetic field. The resulting direct detection demodulation is therefore rather crude when compared with radio or microwave links. Coherent optical links will change this and will begin to exploit the true potential of the optical spectrum. In a coherent system, light is treated as a carrier medium that can be amplitude-, frequency-, or phase-modulated in a manner more analogous to present-day radio systems.

There are three key requirements needed to engineer a coherent optical fiber system.

- There must be a narrow spectral line-width source in which the natural emission spectrum without modulation is narrow compared with the data bandwidth. This also implies frequency stability, such that the laser line width in hertz is less than about 20 percent of the information rate in bits per second.
- The transmission medium must not distort the optical phase front. This condition is readily met in a monomode optical fiber.

- At the receiver, there must be an optical local oscillator laser which is locked to the incoming signal from the remote transmitter. This permits the square law photodetector to be used in either the heterodyne or the homodyne mode (see Section 7.1).

Theoretical studies have shown that up to 20 dB in receiver sensitivity may be obtained by using coherent optical techniques. Obviously, this will result in greatly improved repeater spacing and improvement at high ($>$ 1 Gbit/s) data rates. In this chapter, we discuss the theory of homodyne and heterodyne detection along with techniques for practical realization.

7.1 COHERENT TRANSMISSION TECHNIQUES

A coherent optical transmission link is illustrated in Figure 7.1, along with the spectral characteristics of the transmitter. The dashed lines distinguish the coherent system from the direct detection equivalent. The transmitter consists of a narrow spectral line laser with an external optical modulator that provides amplitude, phase, or frequency modulation. The modulation can be either analog or digital. In some instances, modulation of the laser drive current to produce amplitude or frequency shift keying may be possible.

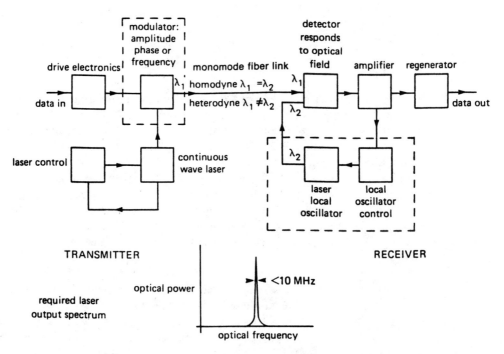

Figure 7.1 Coherent system showing transmitter spectral characteristics. © Copyright 1985 IEEE, IEEE Communications Magazine, Vol. 26, No. 2.

At the receiver, the incoming signal is combined with power from a laser local oscillator in a fused fiber coupler (see Figure 1.27), which provides excellent wavefront matching. The combined signal is then fed to the photodetector. If the two optical wavelengths (frequencies) are identical, the receiver operates in the "homodyne" mode and the signal is recovered directly at baseband. If the local oscillator frequency is offset from that of the incoming signal, "heterodyne" detection is performed. The spectrum of the detector output is centered on an intermediate (IF) frequency the same as a superheterodyne RF receiver. The IF frequency is chosen according to the information rate and modulation characteristics. The operation of this receiver relative to that of a direct detection receiver is illustrated in Figure 7.2.

For both heterodyne and homodyne detection, the reference wave from the

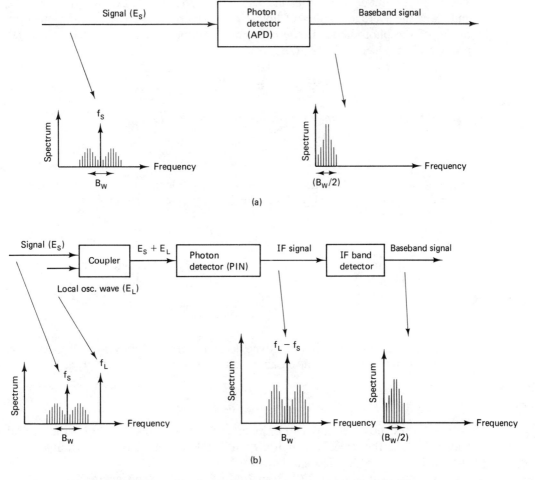

Figure 7.2 Optical receiver block diagram. (a) Direct detection. (b) Heterodyne detection. (From K. Nosu, "Advanced Coherent Lightwave Technologies," *IEEE Communications Magazine,* Feb. 1988. © 1988 IEEE.)

local oscillator laser combines with the incoming signal wave at the photodetector surface. The detector produces an electric current that is proportional to the product of the two incident optical fields. Refer to Equation (2.3). The photocurrent i generated is given by

$$i = \frac{\eta e P}{hf} \qquad (7.1)$$

η = detector quantum efficiency

h = Planck's constant

e = electron charge

The average number of photons incident on the photodiode per unit time is P/hf and i/e is the average number of hole–electron pairs collected across the junction region of the photodiode. If the constant of proportionality that relates optical power and the electric field is $1/Z$, then the power P is given by

$$P = \frac{E^2}{Z} \qquad (7.2)$$

Using this equation, we may express the photocurrent i as

$$i = \frac{\eta e E^2}{Zhf} \qquad (7.3)$$

If the signal carrier field is $E_s = E_{os} \cos W_s t$ and the local oscillator electric field is $E_L = E_{oL} \cos W_L t$, the total peak current $\hat{i}_T$ in the photodetector is

$$\hat{i}_T = \frac{\eta e E_{os} E_{OL}}{hfZ} \cos(\omega_s - \omega_L)t \qquad (7.4)$$

The sum frequency terms are too high to be passed by the detectors. The rms current $\hat{i}_T$ at the IF level is then

$$\hat{i}_{T_1} = \sqrt{\frac{1}{T}\int_0^T \hat{i}_T \, dt} = \frac{\sqrt{2}\eta e E_s E_L}{hfZ} \qquad (7.5)$$

By way of contrast, the homodyne detection process recovers a direct current component from the carrier at baseband and the rms current i_{T_2} is given by

$$i_{T_2} = \frac{2\eta e E_s E_L}{hfZ} \qquad (7.6)$$

In both cases, however, the carrier photocurrent depends linearly on the optical signal field and the effect is to multiply the signal E_s by a factor proportional to E_L, the local oscillator electric field.

We noted in Chapter 2 (Section 2.4.2) that the systems engineer is generally interested in the carrier-to-noise (CNR) performance of the system as a function of the optical input power. For the case of heterodyne detection, the complete expression for the photodetector current i is

$$\begin{aligned} i &= \frac{\eta e}{Zhf}[E_{OS}\cos\omega_s t + E_{OL}\cos\omega_L t]^2 \\ &= \frac{\eta e}{Zhf}\left[\frac{E_{OS}^2 + E_{OL}^2}{2} + E_{OS}E_{OL}\cos(\omega_s - \omega_L)t\right] \\ &= \frac{\eta e}{hf}\left[\frac{(P_s + P_L)}{2} + \sqrt{P_s P_L}\cos(\omega_s - \omega_L)t\right] \end{aligned} \quad (7.7)$$

where the sum frequencies are too high to be passed through the photodetector and can thus be neglected. As noted previously, the $\sqrt{P_s P_L}\cos(\omega_s - \omega_L)t$ term represents the beat between the local oscillator and the signal and is the desirable intermediate frequency (IF) carrier. The IF carrier power depends upon the product of local oscillator power $\sqrt{P_L}$ and the signal light power $\sqrt{P_s}$. Refer to Equation (7.7). In addition to the direct current term and the desired signal current, there is shot noise, which accompanies the direct current. The shot noise power, P_{NS}, is given by

$$\begin{aligned} P_{NS} &= i_{sh}^2 R_L \\ &= 2eR_L B\left[I_D + \frac{\eta e}{2hf}(P_s + P_L)\right] \end{aligned} \quad (7.8)$$

where I_D is the dark current.

The photodetector current also has added to it receiver (circuit) thermal noise P_{NT} which is given by

$$P_{NT} = i_{NT}^2 R_L = 4kTB \quad (7.9)$$

The average signal power P_{ES} delivered to the load resistor is

$$\begin{aligned} P_{ES} &= 0.5 R_L (i_{IF,P})^2 \\ &= 0.5 P_s P_L \left(\frac{\eta e}{hf}\right)^2 \end{aligned} \quad (7.10)$$

where $i_{IF,P}$ is the peak value of the IF signal currents.

The ratio of IF carrier power to noise power CNR is then

$$CN = \frac{P_{ES}}{P_{NS} + P_{NT}} = \frac{0.5 (\eta e/hf)^2 R_L P_s P_L}{2eR_L B[I_D + \eta e/2hf(P_s + P_L)] + 4kTB} \quad (7.11)$$

The local oscillator power can be made much greater than the signal power. In this case, the shot noise will override the thermal noise. Under these conditions, the only noise of consequence is the shot noise produced by the local oscillator. Equation (7.11) reduces to

$$CNR = \frac{\eta P_s}{2hfB} \quad (7.12)$$

Comparing this result to Equation (2.4), we see that in this case the CNR reduces to the quantum limited value. This is the best result obtainable. By raising the power of the local oscillator, we have eliminated the effects of dark current and thermal noise.

Example 7.1

A coherent optical communication system transmits at 1500 nm. The data rate is 100 Mbits/s. Determine the required fractional stability of the laser in hertz.

Solution

$$f_c = \frac{C}{\lambda} = \frac{3 \times 10^8}{1.5 \times 10^{-6}} = 200{,}000 \text{ GHz}$$

$$\text{Fractional stability} = \frac{(0.2)(100 \times 10^6)}{2 \times 10^{14}} = 10^{-7} \text{ parts}$$

Example 7.2

A single-mode fiber optic system consists of a laser diode emitting 10 mW at 1500 nm, a 20-dB fiber cable loss, and a PIN photodetector of responsivity equal to 0.5 A/W. The detector's dark current is 2 nA. The load resistance is 50 ohms, and the receiver's bandwidth is 10 MHz. Its temperature is 27°C. Additional system losses include a 14-dB power reduction due to source coupling. An additional 10-dB loss occurs due to splices and connectors.

(a) Calculate:

- Received optical power
- Detected signal current and power
- The shot noise and thermal noise powers
- The signal-to-noise ratio

(b) Repeat the signal-to-noise ratio calculation if a heterodyne receiver is used and the local oscillator power is $10P_s$.

Solution

(a)
$$\text{System loss} = 14 + 20 + 10 = 40 \text{ dB}$$
$$\text{Received optical power} = P_s - \text{losses}$$
$$= 10 - 44 = -34 \text{ dBm } (0.4 \text{ }\mu\text{W})$$

Using Equation (1.15), we find that the photocurrent is

$$i_s = R_\lambda P = (0.5)(0.4) = 0.2 \text{ }\mu\text{A} = 200 \text{ nA}$$

The 2 nA of dark current can be ignored. The signal power P_s is

$$P_s = i_s^2 R_L = (0.2 \times 10^{-6})^2 (50) = 2 \times 10^{-12} \text{ W}$$

The shot noise power P_{NS} is

$$P_{NS} = i_{sh}^2 \cdot R_L = 2ei_s BR_L$$
$$= 2(1.6 \times 10^{-19})(0.2 \times 10^{-6})(10^7)(50) = 3.2 \times 10^{-17} \text{ W}$$

The thermal noise power, P_{NT} is

$$P_{NT} = i_{NT}^2 R_L = 4kTB$$
$$= 4(1.38 \times 10^{-23})(27 + 273)(10^7) = 1.66 \times 10^{-13} \text{ W}$$

The signal-to-noise ratio is then

$$\text{SNR} = \frac{P_s}{P_{\text{NT}} + P_{\text{NS}}} = \frac{2 \times 10^{-12}}{1.66 \times 10^{-13} + 3.2 \times 10^{-17}}$$

$$= 12(10.8 \text{ dB})$$

(b) Using Equation (7.11), we obtain

$$\text{SNR} = \frac{\eta P_s P_L}{2hf(P_s + P_L)B + i_{\text{eq}}^2 B}$$

In order to solve this equation, the detector quantum efficiency, η is needed. Using Equation (7.1), we have

$$\eta = \frac{ihf}{eP} = \frac{(0.2 \times 10^{-6})(2 \times 10^{-19})/1.5}{(1.6 \times 10^{-19})(0.4 \times 10^{-6})} = 0.417$$

Using Equation (7.11) gives

$$\text{SNR} = \frac{\left[\dfrac{(0.417)(1.6 \times 10^{-19})}{2 \times 10^{-19}/1.5}\right]^2 (50)(0.40 \times 10^{-6})(4 \times 10^{-6})}{2\left[\dfrac{(0.417)(1.6 \times 10^{-19})}{2 \times 10^{-19}1.5}\right](1.6 \times 10^{-19})(4.4 \times 10^{-6})(10^7) + 1.66 \times 10^{-13}}$$

$$= \frac{20.032 \times 10^{-12}}{7.047 \times 10^{-18} + 1.66 \times 10^{-13}} = 120.67(20.8 \text{ dB})$$

7.2 MODULATION TECHNIQUES

Modulation formats for direct detection and coherent optical fiber systems are similar to their counterparts in radio and microwave communication. Coherent transmission offers the choice of phase-, amplitude-, or frequency-shift keying in modulating the optical carrier. The conventional technique of intensity modulation/detection is equivalent to an amplitude-shift-keyed signal directly demodulated by a square law detector. A coherent carrier is not necessary for this approach other than to minimize baseband distortion of the transmitted pulse. This approach places few restraints on the source. As we have seen, however, it does suffer from excessive preamplifier noise. The three basic modulation formats for coherent transmission all assume a source coherence time much longer than a single pulse period. Therefore, the source is treated as a synchronous optical carrier which can be amplitude-, phase-, or frequency-modulated.

The modulation can be effected by using external electro-optic or acousto-optic modulations. In principle, all of the modulation formats that have been developed for radio systems can be used in a coherent optical communication system. However, the criteria for choosing an appropriate format can be quite different for optical systems. As an example, bandwidth efficiency is not an important criterion for an optical system. This eliminates the complexity of such schemes as multilevel

PSK. On the other hand, coherent optical systems suffer from carrier phase noise, which is often a minor consideration in radio and microwave communication.

Frequency modulation of a laser diode can be achieved by changing the laser drive current. The laser drive current determines both the carrier density and the temperature in the semiconductor's active area. These two factors combine to determine the layer's refractive index. The resonant frequency of the cavity depends on its refractive index. Consequently, the resonant frequency (output frequency) changes with the current. Modulation of the drive current then produces frequency modulation of the emitting diode. A circuit for frequency-modulating a laser diode is illustrated in Figure 7.3. The DC current biases the diode to the center of the linear region of its power. This current must be small. For example, a 1-mA change in the current can produce a deviation of 1 GHz in the optical frequency. Large changes in the modulating current will produce undesired large intensity modulation. Electronic limiters are used in the receiver to remove amplitude modulation before demodulating the signal.

For single sine-wave modulation at frequency f_m, the FM frequency deviation is $f = \beta f_m$, where β is the modulation index. If the modulation signal is small, β is linear with the frequency deviation. A representative normalized frequency deviation for a GaAs laser diode is 200 MHz/mA at a modulation frequency of 300 MHz.

Example 7.3

The frequency deviation of a laser diode is 1 GHz/mA at a modulating frequency of 500 MHz. Compute the modulation index when the peak modulating current is 3 mA.

Solution

$$\beta = \frac{\Delta f}{f m} = \frac{3 \times 10^9}{0.5 \times 10^9} = 6$$

External electro-optic and acousto-optic modulators can also be used to effect digital modulation of the laser output. Typical modulation circuits are illustrated in Figure 7.4.

7.2.1 Fiber Requirements

In order for the receiver to work properly, the ideal fiber would maintain a single, linear polarization state throughout its length. At the receiver, the polarization states of the transmitted signal and the local oscillator have to be nearly identical. The

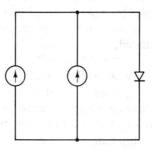

Figure 7.3 Frequency modulation of a laser diode.

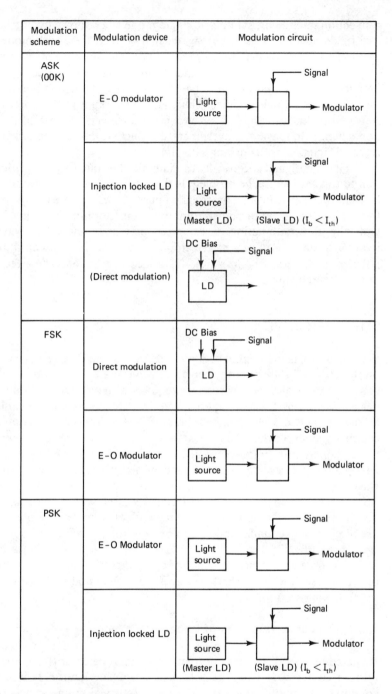

Figure 7.4 Optical transmitters for optical coherent transmission. (From K. Nosu, "Advanced Coherent Lightwave Technologies," *IEEE Communications Magazine*, Feb. 1988. © 1988 IEEE.)

conventional, circular, symmetric single-mode fiber allows two orthogonally polarized fundamental modes to propagate. In a perfect fiber, both modes would propagate together. In practice, however, the fiber contains random manufacturing irregularities that produce geometric and strain-related anisotropic optical effects. The practical effect is a progressive spatial separation of the two polarization modes as they propagate along the fiber. Thus, at a given point in the fiber, the polarization state can be linear, elliptical, or circular. Worse yet, because of mechanical vibration, changes in tension, and temperature fluctuations, the polarization state of the received signal is not constant with time.

Fibers designed specifically to maintain the state of polarization are one solution to this problem. The problem is that the loss is at least double that of the best conventional fiber. From a practical point of view, there are two ways to compensate for the changing state of polarization if a polarization-maintaining fiber is not used. Optically, the combination of a quarter-wave plate and a half-wave plate in fiber retarder or electro-optic devices can potentially realize compensation. Electrically, polarization diversity realizes polarization compensation. These techniques are still in the research stage.

7.3 RECEIVER SENSITIVITY

We noted in Chapter 3 that today's direct detection receivers using APDs requires on the order of 1000 photons per bit to achieve a bit error rate of 10^{-9}. By way of contrast, an ideal coherent receiver requires a signal range of only 10–20 photons per bit to achieve the same bit error rate. A PIN diode may, depending on quantum efficiency, yield one electron per photon. The resulting current is small, and amplification by conventional amplifiers is precluded because of background thermal noise. Coherent techniques promise to replace the noisy APDs and use only stable PIN photodetectors. Because of improved sensitivity, coherent systems may permit wavelength-division-multiplex systems with channel spacings on the order of 100 MHz. This is in contrast with the 1 GHz required with conventional optical multiplexing technology. An added advantage of coherent reception is that it allows the use of electronic equalization to compensate for the effects of optical pulse dispersion in the fiber.

Since we use the performance of direct detection schemes to compare coherent optical systems, we will briefly review the fundamental limits of direct detection in the following section.

7.3.1 Fundamental Limits of Direct Detection

As noted before, direct detection implies a photodetector that converts light energy into an electrical signal. The actual detection mechanism is based upon photon counting. The photon counting process is statistical in nature and is in fact a time-varying Poisson process with intensity function $\lambda(T)$. The mean overall rate of photons at time t is directly proportional to the receiver information data wave.

For binary transmission, the choice between a one and a zero is translated into the presence or absence of a burst of optical energy. Refer to Figure 7.5. The passage of a single pulse through an ideal transmission model is illustrated. When a one is transmitted, the laser or LED is turned on for period T and energy is coupled into the fiber. The transmitted light pulse is detected in the photodetector. The exact time when the photons of light register on the detector is random. The actual electrical current at the output of the photodetector caused by a photon is a wideband pulse, $GW(t)$, where the gain G is either a large integer-valued random variable or $G = 1$, depending upon whether an APD or PIN diode is used. In practical noncoherent systems where amplification of weak signals is required, APDs are invariably used.

If we assume that superposition holds for optical fiber transmission, we can extend the single-pulse description to an entire data wave. Thus, if a sequence of on or off pulses is transmitted, the "received signal," defined as the electrical output of the photodetector on which processing is performed, may be expressed as

$$I(t) = \sum_n G_n W(t - t_n) \tag{7.13}$$

where the random times t_n form a Poisson process with an intensity function $\lambda(t)$ given by

$$\lambda(t) = \sum_n a_n h(t - nt) \tag{7.14}$$

where

$h(t)$ = a square pulse

a = 0 or 1 (data levels)

T = signal interval

G_n = avalanche gain

$W(t)$ = output pulse of the photodetector

With this model of the transmission process, in order to detect the jth bit, the output of the photodetector is integrated over the jth T-second interval. The output signal is a random variable which is then compared with a threshold. If the output is greater than the threshold, a one is declared. If it is less, a zero is declared. With a PIN diode, the gain $G_n = 1$ in the ideal situation and when the threshold is set

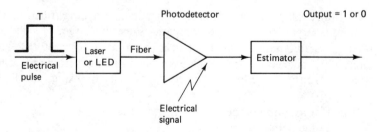

Figure 7.5 On-off direct detection. (From J. Salz, "Modulation and Detection for Coherent Lightwave Communications," *IEEE Communications Magazine,* June, 1986. © 1986.)

at zero, the average output of the integrator will yield $\int_0^T \lambda(t)\, dt = \lambda T$ when a one is sent and a zero when a zero is sent. Since the number of counts n with intensity T is Poisson-distributed, the probability $P(n)$ is given by

$$P(n) = \frac{(\lambda T)^n e^{-\lambda T}}{n!} \tag{7.15}$$

Thus, the probability (chance) of making an error in the detection process is just $\frac{1}{2}P(n)$ ($n = 0$), or

$$P_e = \frac{1}{2} e^{-\lambda T} \tag{7.16}$$

Now, the average optical energy (photons per bit) P is just

$$P = \frac{1}{2}(\lambda T) + \frac{1}{2}(0) \tag{7.17}$$

Using Equation (7.17), Equation (7.16) becomes

$$P_e = \frac{1}{2} e^{-2P} \tag{7.18}$$

This is a fundamental limit on the achievable bit error rate and is commonly referred to as the *quantum limit*.

Example 7.4

Calculate the number of photons per bit needed at the quantum limit for a bit error rate of 10^{-9}.

Solution

$$P_e = \frac{1}{2} e^{-2P} = 10^{-9}$$

$$e^{-2P} = 2 \times 10^{-9}$$

$$P = 10.015 \text{ photons}$$

Note that this limit does not include the effect of any coding. As long as the information rate is less than the channel capacity, the required photons per bit can be reduced by using coding techniques (see Chapter 4).

If an avalanche detector is used to gain optical amplification, the average value of $\{G_n\}$ is large. However, the fluctuations are also large, which results in amplitude jitter. This fluctuation can result in a loss that is 10 to 20 dB from the quantum limit. This, then, is one of the chief motivations for turning to coherent techniques to minimize this large loss in detector sensitivity by using only PIN diodes.

7.3.2 Homodyne Direct Detection and the Super Quantum Limit

Assume that the electromagnetic wave at the output of a laser can be represented as

$$s(t) = A \cos \omega_s t \tag{7.19}$$

where A^2 is proportional to the optical power. Now suppose that the wave is phase modulated so that a one results in $A \cos \omega_s t$ and a zero results in $-A \cos \omega_s t$. As noted in Section 7.1, an ideal homodyne receiver adds to the received wave a local carrier wave with amplitude exactly equal to A. The sum is then

$$S_0(t) = (A \pm A) \cos \omega_s t \tag{7.20}$$

When the sum is detected by a photodetector (PIN diode), the probability of making an error is again described by Equation (7.16). When the output of the photodetector is integrated for T seconds, the average number of counts λT is either $4A^2T$ or 0. The average transmitted optical energy is $P = A^2T$. Thus, using Equation (7.16), we find that the probability of a bit error is

$$P_e = \frac{1}{2} e^{-4P} \tag{7.21}$$

Comparing Equations (7.21) and (7.18), we note a 3-dB improvement over the quantum limit. This is often referred to as the *super quantum* limit. As a practical note, the local laser must know exactly the frequency, phase, and magnitude of the transmitting laser. This is, to say the least, a stringent requirement. This receiver is illustrated in Figure 7.6 with option 1 as the input to the photodetector.

Now in keeping with the discussion of Section 7.1, let us relax the requirements on the local laser and permit its intensity to be any value B. We will still require, however, knowledge of the transmitted carrier frequency and phase. At the receiver, the combined waves become

$$S_0(t) = (B \pm A) \cos \omega_s t \tag{7.22}$$

where B is the amplitude of the local laser and $B \gg A$. The signal described by Equation (7.22) is detected by the photodetector and the average number of counts after integrating over time T is $(B \pm A)^2 T$.

In order to estimate the resulting bit error rate in this situation, a limit theorem from statistics can be applied. This theorem relates to the conditions under which a "shot noise" process can be approximated by a "white Gaussian" noise process. The main requirement is that the rate of photon arrivals be large. Now, since B in

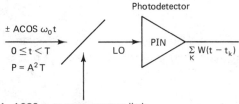

Figure 7.6 Ideal homodyne and heterodyne techniques. (From J. Salz, "Modulation and Detection for Coherent Lightwave Communications," *IEEE Communications Magazine*, June, 1986. © 1986 IEEE.)

Equation (7.22) can be made large (see Example 7.2), the average number of photons is proportional to

$$\lambda T = (B \pm A)^2 T$$
$$= (B^2 + A^2 \pm 2AB)T \qquad (7.23)$$

If we subtract the common bias term $(B^2 + A^2)T$ from λT, we are left with an antipodal signal pair $\pm 2ABT$. This represents the net average counts corresponding to the reception of binary ones and zeros. The variance of the resulting Poisson process is also equal to λT. Now, since $B >> A$, the variance is essentially TB^2. In the limit, then, for a large number of counts due to the addition of the local oscillator to the incoming signal, the photodetector output signal $S(t)$ can be modeled as

$$S_0(t) = \pm 2AB + n(t), \qquad 0 \le t \le T \qquad (7.24)$$

where $n(t)$ is a white Gaussian noise process equal to B^2. Now, if $S_0(t)$ is integrated from 0 to T, the resulting bit error rate is

$$P_e = \frac{1}{2} \text{erfc } \sqrt{2A^2 T} \sim e^{-2A^2 T} \qquad (7.25)$$

where erfc is the complementary error function. This is just the quantum limit and illustrates that an ideal homodyne detector using a PIN diode makes this theoretically possible. Achieving this limit is made possible by the availability of large "local" optical power, which provides indirect amplification of the incoming weak optical signal. This procedure, while providing amplification, also produces additive noise $n(t)$. This mode of detection is illustrated in Figure 7.6 with option 2 as the input to the photodetector. The reader is invited to refer back to the discussion leading to Equation (7.12).

7.3.3 Ideal Heterodyne Detection

Optical heterodyne detection is illustrated in Figure 7.2 and also Figure 7.6 with option 3 as the input to the photodetector. This procedure translates the incoming optical wave to an intermediate (IF) frequency. The local laser frequency is denoted by ω_L. The IF frequency is $\omega_{IF} = \omega_s - \omega_L$. At the photodetector, the addition of the two waves results in

$$S_0(t) = \pm A \cos \omega_s t + B \cos \omega_L t, \qquad 0 \le t \le T \qquad (7.26)$$

where phase modulation is indicated by ± 1 and we require again that $B >> A$. Equation (7.26) can be expressed in terms of the envelope and phase about ω_1. This results in

$$S_0(t) = E(t) \cos [\omega_L T + B(t)] \qquad (7.27)$$

where

$$E(t) = \sqrt{B \pm A(\cos \omega_s t)^2 + A^2 \sin^2 \omega_s t} \qquad (7.28)$$

and

$$B(t) = \tan^{-1} \frac{\pm A \sin \omega_s t}{B \pm A \cos \omega_s t} \qquad (7.29)$$

The response of the photodiode to the wave described by Equation (7.27) is again a shot-noise process with intensity function λ_0 equal to the envelope squared. That is,

$$\lambda_0(t) = B^2 + A^2 \pm 2AB \cos W_s t \qquad (7.30)$$

Using the same limit arguments as in Section 7.3.2, we first subtract the bias term $A^2 + B^2$ from Equation (7.30). This yields the antipodal signal pair $\pm 2AB \cos W_s t$. Since $B >> A$, the fluctuating noise is white Gaussian with double-sided spectral density $\sim B^2$. If we again denote the additive noise by $n(t)$, the equivalent signal plus noise may be expressed as

$$S_0(t) = \pm 2 AB \cos \omega_s t + n(t) \qquad 0 \le t \le T \qquad (7.31)$$

In the literature, this is a standard detection problem and deciding whether a plus or minus was sent is accomplished by multiplying $S(t)$ by $\cos \omega_s t$ and integrating for T seconds. The result of this integration is compared to a threshold set to zero. The decision statistic is

$$\int_0^T S_0(t) \cos \omega_s t \, dt = \int_0^T [\pm 2 AB \cos \omega_s t + n(t)] \cos \omega_s T \, dt \qquad (7.32)$$
$$= \pm 2 ABT + \int_0^T n(t) \cos \omega_s T \, dt$$

where the twice frequency term is neglected (filtered). Since the random variable $\int_0^T n(t) \cos \omega_S t \, dt$ has variance equal to $B^2 T/2$, the bit error rate in this case is asymptotically equal to

$$P_e \approx e^{-A^2 B^2 T / B^2 T} = e^{-A^2 T} = e^{-P} \qquad (7.33)$$

Comparing this expression to Equation (7.25), we see that heterodyne detection is 3 dB (factor of 2) better than the quantum limit. The performance of the detection schemes discussed is summarized in Table 7.1. The values tabulated in Table 7.1 represent ideal performance. In more realistic systems, it is necessary to take laser phase noise into consideration. Phase noise is discussed in Section 7.3.4.

7.3.4 Phase Noise in Lasers

We noted in the introduction to this chapter that a coherent optical system requires a narrow spectral line-width source. The fundamental limit on the performance of

TABLE 7.1 IDEAL PERFORMANCE

1. Super homodyne	e^{-4P}
2. Homodyne	e^{-2P}
3. Heterodyne	e^{-P}
$P = A^2 T$ = energy/bit	

a coherent optical system is set by the phase or frequency noise in the laser. The spectral density of this noise exhibits a $1/f$ to $1/f^2$ characteristic up to around 1 MHz and is flat for frequencies above 1 MHz (see Figure 7.7). The "white" or flat component results from quantum fluctuation and is the principal cause of line broadening. From a communications theory point of view, the relatively low-frequency components can be easily tracked, leaving the white noise the primary problem area. The physical source of laser phase noise is the randomly occurring spontaneous emission events. Each event causes a sudden jump of the magnitude and sign of the device's output electromagnetic (EM) field. Then as time elapses, the phase of the EM field executes a random walk away from the value it would have had in the absence of spontaneous emission. It is shown in the literature that the mean squared value of the random phase, $\theta(t)$, becomes in the limit a Weiner process characterized by a zero-mean white Gaussian frequency noise, $n(t)$, with two-sided noise spectral density N_0. Thus, the phase process can be represented as

$$\theta(\tau) = 2\pi \int_0^\tau \dot{\theta}(t)\, dt \tag{7.34}$$

and the mean squared phase deviation is given by

$$E[\theta^2(\tau)] = E\left[2\pi \int_0^\tau \dot{\theta}(t)\, dt\right]^2 = (2\pi)^2 N_0 \tau \tag{7.35}$$

One scheme for determining N_0 makes use of the fact that phase noise causes an observable broadening of the laser emission spectrum. Consider the sine-wave random process

$$S_0(t) = A \cos\left[2\pi f_0 t + \theta(t) + \phi\right] \tag{7.36}$$

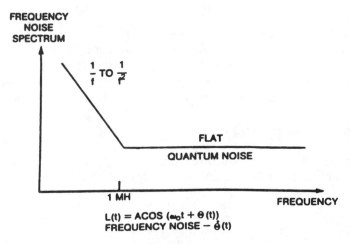

Figure 7.7 Laser phase noise. (From J. Salz, "Modulation and Detection for Coherent Lightwave Communications," *IEEE Communications Magazine*, June, 1986. © 1986 IEEE.)

The inclusion of the uniform phase ϕ makes $S_0(t)$ a stationary process with the following correlation function:

$$R(\tau) = E[S_0(t)S_0(t+\tau)]$$
$$= \frac{A^2}{2} Re\left\{ e^{j2\pi f_0 \tau} - \frac{(2\pi)^2}{2} N_0 \tau \right\} \quad (7.37)$$

The Fourier transform of this expression yields the power spectrum $G(f)$. That is,

$$G(f) = \frac{A^2}{4\pi^2 N_0} \left\{ \left[1 + \left(\frac{f+f_0}{\pi N_0}\right)^2 \right]^{-1} + \left[1 + \left(\frac{f-f_0}{\pi N_0}\right)^2 \right]^{-1} \right\} \quad (7.38)$$

A sketch of the baseband power spectrum commonly called the "Lorentzian" is shown in Figure 7.8. The -3-dB bandwidth is denoted as B_L. From Equation (7.38) we see that for $f_0 \gg B_L$, the noise density N_0 is equal to

$$N_0 = \frac{B_L}{2\pi} \quad (7.39)$$

The quantity B_L is commonly referred to as the laser line width.

We noted in the introduction that the laser line width must be less than 20 percent of the information rate in bits per second (also see Example 7.1). For comparison, note that microwave oscillators, which are widely used in coherent radio applications, have line widths on the order of 1 Hz. In order to reduce laser line width, experimenters have exploited the fact that the noise density N_0 is inversely proportional to $P_0 Q^2$, where P_0 is the laser output power and Q is the quality-factor of the "cold" laser cavity resonance; thus, high-power, high-Q lasers tend to have

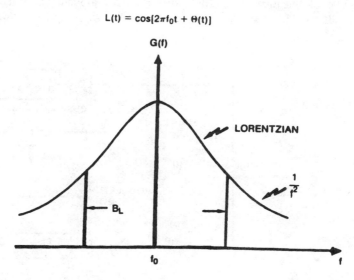

Figure 7.8 Power spectrum of laser line. (From J. Salz, "Modulation and Detection for Coherent Lightwave Communications," *IEEE Communications Magazine*, June, 1986. © 1986 IEEE.)

narrow line widths. As of this writing, under laboratory conditions, line widths on the order of tens of kHz have been obtained.

7.3.4 Phase-Lock Techniques

We saw in Section 7.3.3 that homodyne detection makes it theoretically possible to achieve the quantum limit. We deliberately avoided a discussion at that time of the practical aspects surrounding the local oscillator. In order to realize the full benefits of the homodyne approach, the local laser must have perfect knowledge of the transmitted optical center frequency and phase. These crucial parameters must be derived directly from the optical wave or from a heterodyned version. Carrier recovery techniques at microwave frequencies are well defined. At optical frequencies the optical counterparts are still limited.

One scheme for estimating or tracking the phase noise of the incoming optical signal so that it can be used to coherently demodulate PSK is illustrated in Figure 7.9. Detection is accomplished directly by using only the intensity of the optical wave. At the input to the detector, the power of the incoming optical data signal is split so that a fraction, A^2k^2, is directed to the phase-locked loop. The remaining portion of the power, $A^2(1 - k^2)$, is used for demodulation. The division of the power is determined by k ($0 \le k \le 1$), and the choice of this constant must be optimized. As of this writing, no experimental verification of this detection scheme has been found in the literature. A variation that has been proven experimentally is illustrated in Figure 7.10.

The significant features in this design are a twin photodiode receiver and a balanced phase-locked loop. This arrangement is used to reduce the effect of optical fluctuations from the local oscillator phase. The difference signal from the two detectors of the local oscillator controls mirror movement by using a piezoelectric transducer. This design proved capable of phase-locking a microwatt of local oscilla-

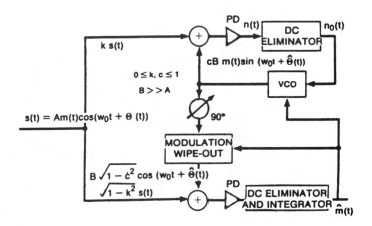

Figure 7.9 Optical homodyne detector. (From J. Salz, "Modulation and Detection for Coherent Lightwave Communications," *IEEE Communications Magazine,* June, 1986. © 1986 IEEE.)

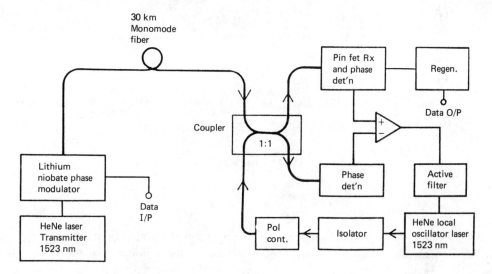

Figure 7.10 Homodyne system. (From I.W. Stanley, "A Tutorial Review of Techniques for Coherent Optical Fiber Transmissions Systems," *IEEE Communictions Magazine*, Aug. 1985. © 1985 IEEE.)

tor power to a signal level three orders of magnitude less. Heterodyne techniques, while offering 3 dB less in performance, ease the requirements for phase-locked loops. This is discussed in the following section.

7.3.4.1 Heterodyne-Phase-Locked Loops. Using heterodyne techniques, the optical data wave is moved down to an IF frequency first and the carrier is then derived from the resulting microwave signal by using well-known standard techniques. Heterodyning the optical signal to an IF frequency, f_i, and wiping off the modulation result in a microwave signal plus noise described by

$$S(t) = 2A \cos[\omega_i t + \theta(t)] + n(t) \tag{7.40}$$

where

A^2 = optical power

$n(t)$ = white Gaussian noise with unit double-sided spectral density

$\theta(t)$ = difference between the transmitting laser's phase noise and the local laser's phase noise

The signal represented by Equation (7.40) is now the input to a conventional PLL, as illustrated in Figure 7.11. We previously indicated that the performance of heterodyning is 3 dB inferior to the homodyne case (see Table 7.1). The additional degradation is expected to result from the nonperfect estimation of the laser phase noise. A practical heterodyne system employing PSK modulation is illustrated in Figure 7.12. The optical sources are helium–neon gas lasers, which exhibit narrow spectral line widths. With PSK, no carrier component is transmitted, so it is necessary to

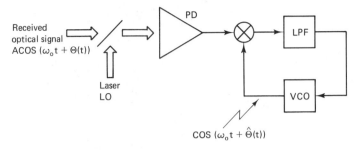

$\Theta(t) - \hat{\Theta}(t) = \Psi(t)$ — phase error process

Figure 7.11 Phase-lock techniques.

reinsert it at the receiver. The arrangement illustrated consists of a squaring and divide-by-two circuit with the phase-locked loop. For differential PSK, this carrier synchronization circuit would be replaced by a path delay demodulation with a time shift of one bit period. In order to maintain the 210-MHz IF frequency accurate to within 1.5 MHz, an AFC control scheme is used, which includes a frequency discriminator that alters the optical path length in the long external cavity laser by rotating a silica plate.

The theoretical performance limits for the most commonly employed modulation techniques are tabulated in Table 7.2.

The minimum detectable (peak) power level for a 10^{-9}-bit error rate is obtained by multiplying the number of photons with hfB/η, where B is the bit rate, η is the

TABLE 7.2 SENSITIVITY OF OPTICAL RECEIVERS

Modulation/Detection Type	BER	Number of Photons for BER of 10^{-9}, $\eta = 1$
ASK heterodyne	$0.5 \, \text{erfc} \left(\sqrt{\dfrac{\eta P_s}{4hfB}} \right)$	72
ASK homodyne	$0.5 \, \text{erfc} \left(\sqrt{\dfrac{\eta P_s}{2hfB}} \right)$	36
FSK heterodyne	$0.5 \, \text{erfc} \left(\sqrt{\dfrac{\eta P_s}{2hfB}} \right)$	36
PSK heterodyne	$0.5 \, \text{erfc} \left(\sqrt{\dfrac{\eta P_s}{hfB}} \right)$	18
PSK homodyne		9
Direct detection	$0.5 \, \text{erfc} \left(\sqrt{\dfrac{2\eta P_s}{hfB}} \right)$	21
Quantum limit practical receiver	$0.5 \, \text{exp} \left(\sqrt{\dfrac{-\eta P_s}{hfB}} \right)$	400–4000

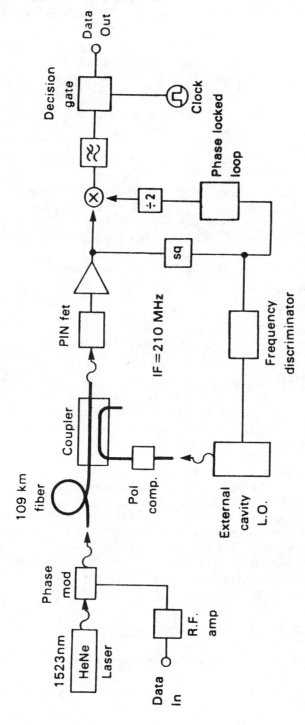

Figure 7.12 Arrangement of components for a 140 mb's heterodyne PSK transmission test. (From I.W. Stanley, "A Tutorial Review of Techniques for Coherent Optical Fiber Transmission Systems," *IEEE Communications Magazine*, Aug. 1985. © 1985 IEEE.)

detector quantum efficiency, hf is the energy of a photon, and P_s is the peak signal power of the received signal. Note from this table that PSK homodyne is the most sensitive of the binary coherent schemes. When the promised performance of coherent systems is contrasted with that of direct detection systems, the rationale for today's research in this area is vividly demonstrated.

REFERENCES

7.1. SALZ, J. 1986. Modulation and detection for coherent lightwave communications. *IEEE Communications Magazine,* June, 38–49.
7.2. URY, ISRAEL. 1985. Optical communications. *Microwave Journal,* April, 24–35.
7.3. BASCH, E. E., and BROWN, T. G. 1985. Introduction to coherent optical fiber transmission. *IEEE Communications Magazine* 23:23–29.
7.4. STANLEY, I. W. 1985. A tutorial review of techniques for coherent optical fiber transmission systems. *IEEE Communications Magazine* 23:37–53.
7.5. NOSU, KIYOSHI. 1988. Advanced coherent lightwave technologies. *IEEE Communications Magazine* 26:15–21.

PROBLEMS

7.1. A heterodyne system employs an RF filter following photodetection. The filter is tuned for a center frequency of 1 GHz with a 10-MHz bandwidth. The local oscillator operates at a frequency of 10^{14} Hz, and the received carrier is at an optical frequency such that mixing with the local oscillator produces the proper RF for the filter.
  (a) How much optical shift of the received carrier is acceptable in the system? Neglect carrier modulation.
  (b) How much local oscillator instability can occur?

7.2. The laser diode in Example 7.2 couples 2 mW of power into the fiber.
  (a) How much local signal power is needed to achieve a strong oscillator condition, assuming that a 20-dB local to received field power ratio is needed?
  (b) How much is needed to overcome circuit noise with the same ratio as (a) when the detector load resistance is 100 ohms and operates at room temperature?
  (c) If the detector has gain G, determine how the answers in (a) and (b) are affected.

7.3. The linear loop in Problem 7.4 has a loop transfer function from loop input to VCO output given by

$$H(s) = \frac{G/s}{1 + (G/s)}$$

where G is the total loop gain. Show that the loop bandwidth B_L is given by

$$B_L = \frac{1}{2\pi} \int_0^\infty |H(\omega)|^2 \, d\omega = \frac{G}{4}$$

7.4. Use the linear loop model in Figure P7.4.
   (a) Show that the transform of the phase error is related to that of the phase input by $F_\phi(s) = F_{\theta 1}(s)H(s)$, where

$$H(s) = \frac{G/s}{1 + (G/s)}$$

   (b) Use (a) to determine the error response when $\theta_1(t) = \theta_0$.
   (c) Repeat (b) when $\theta_1(t) = \omega_0 t$.
   (d) Repeat (c) when $\theta_1(t) = a \sin \omega_0 t$.

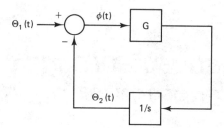

Figure P7.4

7.5. Repeat Example 7.2 if an APD with a gain of 100 is used.

7.6. Show that $\frac{1}{2}\text{erfc}(2A^2T)^{1/2} \approx e^{-2A^2T}$

7.7. The maximum allowable drift rate of a coherent communication system is specified in terms of parts per day, where the term *parts* refers to the fractional frequency error of the oscillator. For example, a 1-GHz oscillator with an error of 1 Hz has a fractional error of $1/10^9 = 10^{-9}$ parts, or hertz per hertz per day. If the total allowable error excursion is ± 500 sec in 30 days, find the maximum allowable drift rate for a reference oscillator in parts per day. Assume that the oscillator frequency is 10^{14} Hz.

7.8. (a) Show that Equation (7.38) is true.
   (b) Derive an expression for the half-power frequencies, beginning with Equation (7.37).

7.9. (a) Find the resonant frequency for a laser that corresponds to a wavelength of 1 μm.
   (b) If the line width is 1.884×10^9 rad/s and the resistance R is 5.313×10^{-10} ohms, what is the equivalent inductance L? Is it possible to generate an inductance of this value?
   (c) Express the line width in wavelengths.
   (d) Calculate the equivalent capacitance.

7.10. Find the Fourier transform of Equation (7.36).

8

Measurements in Fiber Telecommunications

8.0 INTRODUCTION

Fiber transmission systems have been in existence now for a number of years. Generally speaking, these systems have been maintained by elite groups of specialized personnel. However, since the divestiture of AT&T, along with the advent of single-mode fiber, a transition has occurred. Today, fiber transmission systems are being installed worldwide at an astonishing pace. Consequently, attention is turning to the maintenance of installed systems. Maintaining fiber transmission systems requires new test technologies along with trained personnel and equipment.

Measurements on fiber cable installation can be divided into four groups:

- Cable continuity
- Cable attenuation or loss
- Optical time domain reflectometry
- Cable dispersion or bandwidth

In a manner similar to radio systems, there are several basic tools used for test and analylsis of fiber optic systems. Test equipment used on fiber systems can be divided into six groups. These are

- Optical power
- Optical attenuation
- Optical time domain reflectometry
- Optical bandwidth
- Optical margin
- Signal/noise (for analog systems) and bit error rate (for digital systems)

A discussion of these cable and system tests is presented in the following.

8.1 FIBER CABLE MEASUREMENTS

Cable continuity is the easiest test, requiring only a simple right optical light source and an optical power meter (Figure 8.1). For shorter lengths, a flashlight and the naked eye are sometimes used. This is not recommended, however, for reasons of eye safety. Continuity provides a Go–NoGo qualitative test. Either the fiber cable passes light or it doesn't. No attempt is made to measure the actual loss of the fiber. Thus, neither the optical source nor the power meter requires stability and accuracy.

8.1.1 Spectral Attenuation of Fibers

Cable attenuation, generally referred to as *cable loss*, is a more elaborate version of cable continuity. For this measurement, we again require an optical source and a power meter. In this case, however, stability and accuracy are very relevant. The power meter must be accurate, and the optical source must be stable over both time and temperature. Often the combination of a stable optical source and power meter is referred to as an *optical loss set*. The loss set may be packaged as a single instrument or may remain as two separate units.

Attenuation is defined as the ratio of light power coupled into the fiber cable to the light power coupled out of the cable's far end, expressed in decibels. The attenuation of any fiber is wavelength-dependent. A plot of loss versus wavelength is referred to as a spectral loss curve. The attenuation of the fiber cable should be measured, then, at the wavelength of interest, hence the requirement for source stability.

Attenuation measurements on fiber cables and systems can be accomplished in several ways. Refer to Figure 8.2. The loss testing illustrated in Figure 8.2(a) uses the combination of a stable optical source and a power meter. This technique has the advantage of being easy to perform. The disadvantage is that an identical reference fiber must be used, or one fiber must have a piece cut off (called the *cut-back*

Figure 8.1 Cable continuity check.

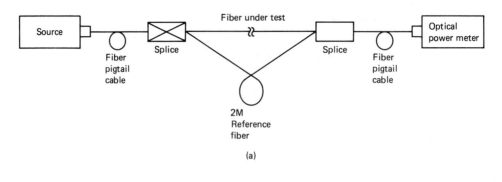

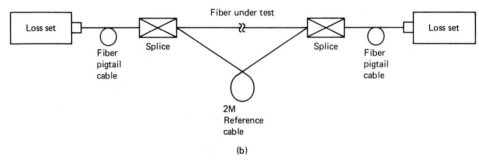

Figure 8.2 (a) Loss testing with a stable source and power meter using reference fiber substitution. (b) Loss testing with two loss test sets using reference fiber substitution.

method). In this technique, power emanating from two lengths of fiber, or one length of fiber before and after a cut, is measured at the desired wavelength. The cable loss in dB/km is

$$\text{Loss} = \frac{P_1 - P_2}{L} \qquad (8.1)$$

where

P_1 = power out of cable 1, dB

P_2 = power out of cable 2, dB

L = difference in cable lengths, km

This measurement can be repeated over a sequence of wavelengths to determine the spectral attenuation. The setup illustrated in Figure 8.2(b) illustrates the same test performed with optical loss sets. The difference here is that loss can be measured in both directions with little additional effort.

End-to-end attenuation measurements can also be made on installed systems. In the case of an installed system, the multifiber cable plant is typically fanned out into individual jacks on optical patch panels. Connecting an optical source (or loss set) to one end and a power meter (or loss set) to the far end will provide the cable plant's attenuation. This test is illustrated in Figure 8.3.

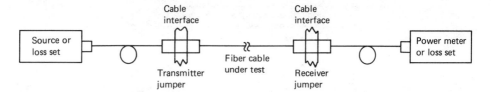

Figure 8.3 Cable plant attenuation testing with source and power meter or loss sets.

8.1.2 Optical Time-Domain Reflectometer

Another basic method used to measure attenuation is the *backscatter* technique. In the backscatter technique, a high-intensity, short-duration light pulse in the range of 20–2000 ns is launched into an optical fiber. An oscilloscope detects, amplifies, and displays returning reflections. The return reflections arise from Rayleigh scattering in the fiber and discontinuities at connectors, splices, and breaks. The optical time-domain reflectometer (OTDR) performs this backscatter measurement. Refer to Figure 8.4. The OTDR functions as a one-dimensional optical radar. A light pulse from the source is launched into the fiber. The beamsplitter (directional optical coupler) prevents the initial laser light from reaching the photodetector. The photodetector records only the returned signals (backscattered light). An idealized plot of an OTDR signal display (signature) is shown in Figure 8.5.

The joints, faults, or discontinuities in the fiber can be located by noting the rise or fall of signal level within 1 meter (about 10-ns time resolution). Two distinct features are present on the signature. First, reflections arise from discontinuities such as small bubbles in the fiber and Fresnel reflections from couplers, breaks, and certain types of splices. Second, a continuous return is observed, which is due to Rayleigh scattering of the probe pulse traveling in the forward direction. The ob-

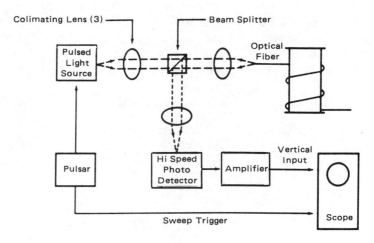

Figure 8.4 Basic configuration of the OTDR.

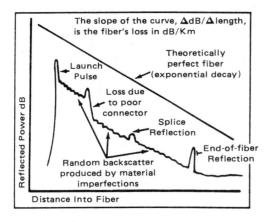

Figure 8.5 Representative signal from an OTDR (signature).

served signal represents the two-way travel time through the device under test. The fiber's loss in dB/km is given by the slope of the curve.

8.1.3 Dispersion Measurements

Attenuation along a fiber that is frequency-dependent results in a dispersed signal. Visually, the effect on a pulse is a broadening of the pulse with respect to time. Dispersion was discussed in Chapter 3 (see Section 3.2.4). Essentially, some modes take longer to traverse a fiber than others. Thus, modes that are launched simultaneously do not arrive at the opposite end of the fiber at the same time. Since the observed modal delay depends on the specific modes present throughout the length of the fiber, the observed delay will depend on a variety of conditions. Specifically, these are (1) the actual modes launched by the source, (2) mode mixing, and (3) mode attenuation. It is difficult, then, to extrapolate modal delay from one length of fiber to another or from one set of launch conditions to another. We know from the discussion in Chapter 3 that in addition to modal dispersion, chromatic dispersion results from both material dispersion (different wavelengths propagating at different effective velocities) and waveguide dispersion of an individual mode. For multimode fibers, modal dispersion is usually the dominant effect. We can calculate these two types of dispersion using the following relations:

$$\delta_m = \frac{L(\text{NA})^2}{2N_c C} \tag{8.2}$$

$$\delta_{\text{mat}} = \left(\frac{L\lambda_s \lambda}{c}\right) \frac{d^2 N}{d\lambda^2} \tag{8.3}$$

$$\delta_c = \delta_{\text{mat}} + \delta_w \tag{8.4}$$

$$\delta_{\text{tot}} = \sqrt{\delta_c^2 + \delta_m^2} \tag{8.5}$$

where

δ_m = modal dispersion

δ_{mat} = material dispersion

δ_w = waveguide dispersion

δ_c = chromatic dispersion

δ_{tot} = total dispersion

L = cable length

c = velocity of light in vaccum

λ_s = 3-dB (power) bandwidth of source

N_c = effective refractive index

NA = numerical aperture

The chromatic dispersion δ_c can be determined by measuring the input pulse width W_{in} and the output pulse width W_{out}. The chromatic pulse width is calculated as

$$\delta_c = \Delta t = \sqrt{W_{out}^2 - W_{in}^2} \text{ s} \tag{8.6}$$

Since a short-duration light pulse is applied, W_{in} and W_{out} can be almost equal. A delay measurement can be utilized to accentuate the pulse broadening (see Figure 8.6). From the oscilloscope display, the pulse width D_1 with one foot of cable is measured. The same measurement is then repeated with the one-foot cable replaced by a one-km cable. The chromatic dispersion is then

$$\delta_c = D = D_2 - D_1 \text{ s/km} \tag{8.7}$$

Using Equations (8.7), (8.2), and (8.5), we find that the total dispersion is given by

$$\delta_{tot} = \sqrt{D^2 + \left[\frac{L(NA)^2}{2N_c C}\right]^2} \tag{8.8}$$

In order to make use of Equation (8.8), it is necessary to known the numerical aperture NA and the "effective refractive index." The OTDR permits the quantitites to be determined.

8.1.3.1 Measurement of Refractive Index Profiles.

The OTDR may be used to measure the effective refractive index and numerical aperture of a fiber optic cable. We noted previously that the OTDR is essentially a one-dimensional, closed-circuit, optical radar, requiring the use of one end of a fiber to make measurements. The measurements are attenuation at breaks, connectors, and other fiber imperfection points. This data is obtained from the input parameters, that is, the pulse width and the refractive index set into the OTDR. Because of the relationship of the length and refractive index, the cable length displayed by the OTDR is accurate only if the refractive index set into the OTDR is accurate.

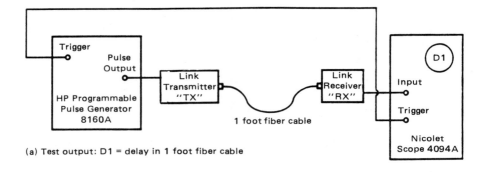

(a) Test output: D1 = delay in 1 foot fiber cable

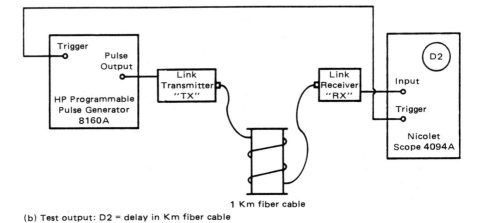

(b) Test output: D2 = delay in Km fiber cable

Figure 8.6 Experimental setup for measuring chromatic dispersion. (a) Test output: D1 = delay in 1-foot fiber cable. (b) Test output: D2 = delay in 1-km fiber cable.

After determining the travel time of a light pulse, the OTDR calculates distance D according to Equation (8.9)

$$D = \frac{ct}{2N} \qquad (8.9)$$

where

D = distance to the defect or end of the fiber

t = round-trip travel time between the launched pulse and the returned pulse

c = speed of light in a vaccum

N = average refractive index (group index) of the optical fiber core

If the effective refractive index of the optical fiber within the cable does not match the core refractive index in the specification, the distance measured to a break can be quite inaccurate. If this is the case, Equation (8.10) can be used to calculate the effective refractive index.

$$N_c = \frac{N_f D_2}{D_1} \tag{8.10}$$

where

N_c = correct effective refractive index

N_f = the index of refraction set into an OTDR, which then indicates an optical fiber length of D_2

D_1 = physically measured cable length

D_2 = cable length indicated by the OTDR with refractive index N_f set into it

The numerical aperture can be calculated as follows:

$$\text{NA} = N_c \sqrt{2\Delta} \tag{8.11}$$

where

$$\Delta = \frac{N_c - N_{c_1}}{N_{c_1}}$$

NA = numerical aperture

N_{c_1} = cladding refractive index

Δ = refractive index difference

8.2 FIBER BANDWIDTH MEASUREMENTS

Optical loss (attenuation) and bandwidth are two fiber transmission parameters of primary interest when fiber systems are designed. Fiber bandwidth limits the maximum rate at which information can be transmitted over a fiber. We know from previous discussions that signal degradation resulting from such factors as dispersion causes the pulse width to spread. If this spreading becomes too large, the broadened pulse can interfere with pulses on either side of it, resulting in intersymbol interference along with high bit error rates.

The wider the spread with time is, the narrower the bandwidth. This description has intuitive appeal and is frequently referred to as the "time domain" description of bandwidth. Many bandwidth test sets use the time domain technique-measuring input and output pulses and then mathematically transform these pulses into a "frequency domain" description from which bandwidth is determined. The word *bandwidth* is a frequency-domain term and is inversely proportional to its time-domain counterpart, *pulse spreading*.

8.2.1 Comparison of Time- and Frequency-Domain Measurements

Time- and frequency-domain methods for measuring bandwidth are mathematically equivalent and should, therefore, provide the same result. A measurement of bandwidth in the frequency domain is determined directly from the measured response curve (see Figure 8.7). Most field test sets use the frequency-domain technique. By way of contrast, in the time domain this curve is obtained mathematically by dividing the Fourier transform of the output pulse by the Fourier transform of the input pulse. For either technique, the measured bandwidth can depend markedly on the wavelength of the light source, the spectral width of the source (laser or LED), and the launch conditions. The bandwidth can vary, for example, depending on whether the light uniformly illuminates the fiber cross section or simply illuminates a small spot.

We discussed in previous chapters the various mechanisms that limit a fiber's bandwidth by causing pulses of light to arrive at the end of the fiber at slightly different intervals in time. These bandwidth-limiting mechanisms can be divided into the following categories:

- Modal (sometimes called *intermodal*)
- Chromatic (sometimes called *intramodal*)

The relative importance of these two mechanisms in limiting the maximum bandwidth of the fiber depends on the characteristics of the light source and whether the fiber is single mode or multimode. The relative importance of these bandwidth-limiting mechanisms is illustrated in Table 8.1.

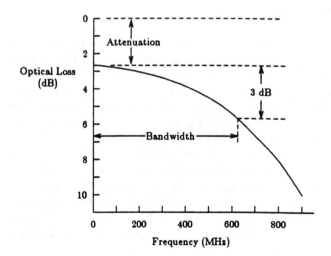

Figure 8.7 Optical loss of a fiber as a function of modulating frequency.

TABLE 8.1 MECHANISMS LIMITING FIBER BANDWIDTH

	Fiber Type	
Source	Multimode	Single-Mode
Laser	Modal	Chromatic
LED	Modal and chromatic	Chromatic

8.2.2 Bandwidth Specification by Fiber Manufacturers

Manufacturers measure the bandwidths of multimode fibers using laser sources having wavelengths at 825 nm (or 850 nm) and/or 1300 nm. Bandwidth measurement is performed during the fiber-spool stage of manufacture before the fibers are jacketed or placed into cable structures. At this stage, the fiber is wound on a plastic spool (approximately 6-inch hub diameter) and has a length of 2 km or more. The bandwidth reported by the manufacturer is in MHz·km. Basically, the manufacturer multiplies the measured bandwidth times the length of the fiber. This MHz·km value is a figure of merit that can be used to compare the bandwidth quality of fibers available from a given manufacturer or from different manufacturers.

It should be noted that the bandwidth of an installed lightwave transmission system depends on the final exit bandwidth of the installed and spliced fiber section when measured with that system's source—not the fiber manufacturer's source. If the operating wavelength of the system source (either LED or laser) is similar to that used by the manufacturer, then the bandwidth of the installed section or system can be related to the manufacturers' spooled-fiber bandwidth.

Note from Table 8.1 that the bandwidth-limiting mechanism for multimode fiber when used with a laser source is modal bandwidth. We will designate the bandwidth measured with a laser as $BW_{s(laser)}$. If the wavelength of the laser is similar to that of the fiber manufacturer, the bandwidth can be correlated to the manufacturer's spooled-fiber bandwidth as follows:

$$BW_{s(laser)}(MHz) = BW_f(MHz \cdot km) l_s^{-\gamma_{laser}} \qquad (8.12)$$

where

$BW_{s(laser)}$ = system bandwidth in MHz measured with a laser

BW_f = spooled-fiber bandwidth in MHz·km measured with a similar laser

l_s = installed section (system) length in km

γ_{laser} = laser bandwidth concatenation length scaling factor

γ_{laser} typically has a value between 0.5 and 1.0. It depends on the system's operating wavelength and how the fibers were designed, manufactured, and spliced. As an example, if the system operating wavelength differs substantially from the wavelength for which the fiber has the highest bandwidth, the value for γ_{laser} will be near unity. Conversely, if the system wavelength coincides with this wavelength, γ_{laser} will be closer to 0.5.

A comprehensive equation for bandwidth takes into account packaging, installation, and length scaling effects. This is beyond the scope of the simple "gamma" equation (8.12). Based on laboratory and field measurements, the following equation was developed for the laser bandwidth multimode fiber.

$$BW_{s(laser)}(MHz) = aBW_f(MHz \cdot km) l_s^{-\gamma_{laser}} l_c^{-b} \tag{8.13}$$

where

$BW_{s(laser)}$ = laser section bandwidth in MHz

a = a factor to account for bandwidth changes when the cables are unreeled and installed

BW_f = manufacturer's spooled-fiber bandwidth in MHz·km

l_s = installed section length in km

l_c = average cable length in km

γ_{laser} = laser bandwidth concatenation length scaling factor

b = factor to account for differences between concatenation and cut-back length scaling factors

The b term in Equation (8.13) can be understood by considering what happens when 1 km of fiber is cut from a 3-km spool whose measured bandwidth is 300 MHz (the spool has 900-MHz·km bandwidth). Now, if three such 1-km lengths (each spool having 900-MHz·km bandwidth) were selected randomly and spliced together, the section bandwidth of the 3-km spliced length would be closer to 400 MHz than to the original 300 MHz of the unspliced lengths. The higher bandwidth occurs because the length scaling factor for randomly selected fibers that are spliced together (the concatenation γ) is smaller than for a uniform fiber being cut into shorter pieces (the cut-back γ). In summary, the bandwidths of many short pieces of fiber scale more favorably than the bandwidths of a few long pieces.

8.2.2.1 LED-Based Multimode Fiber Systems. Table 8.1 illustrates that a multimode fiber with LED source exhibits both modal (laser) and chromatic effects. Thus, the bandwidth, $BW_{s(LED)}$, of a section measured with an LED consists of both the laser bandwidth plus chromatic bandwidth. Chromatic bandwidth depends on the range of wavelengths contained in the source, together with their spectral location. Three parameters are frequently used to describe the shape of an LED's optical power spectrum. Refer to Figure 8.8. These parameters are (1) the peak wavelength λ_p, where the output power peaks, (2) the spectral width of the source λ_w, which is the width of the source at its half-maximum amplitude points, and (3) the center wavelength λ_c, the average of λ_1 and λ_2.

If the LEDs power spectrum is symmetrical about λ_p, then $\lambda_p = \lambda_c$. Frequently, this is not true and λ_p is difficult to determine because of undulation on the spectral curve. For these reasons we use λ_c instead of λ_p in the equations for chromatic

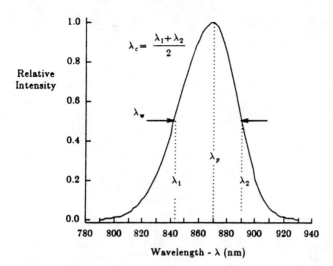

Figure 8.8 Three parameters are frequently used to describe the location and width of a power spectrum. Shown here for an LED are peak wavelength, center wavelength, and spectral width.

bandwidth. Equations for predicting the chromatic bandwidth of multimode fibers are tabulated in Table 8.2. Since both chromatic and laser bandwidths combine to form the total bandwidth of a section measured with an LED, the total bandwidth is given by

$$BW_{s(LED)} = \frac{BW_{s(chrom)} BW_{s(laser)}}{\sqrt{BW_{s(chrom)}^2 + BW_{s(laser)}^2}} \qquad (8.14)$$

TABLE 8.2 INSTALLED SECTION CHROMATIC BANDWIDTH $BW_{s_{chrom}}$ (MHz)

Wavelength (nm)	
Short (near 870 nm)	Long (near 1300 nm)
$\dfrac{k\lambda_c l_s^{-\gamma_{chrom}}}{|a_1 \lambda_c^4 - a_2|\lambda_w}$	$\dfrac{l_s^{-\gamma_{chrom}}}{\sqrt{\lambda_w}(b_1 + b_2|\lambda_c - b_3|)}$

λ_c = center wavelength of the LED in nm

γ_{chrom} = chromatic bandwidth length scaling factor

λ_w = full-width, half-maximum spectral width of LED in nm

l_s = section length in km

k, a_i, and b_i = constants that depend on wavelength and fiber type

This equation can be used with the equations in Table 8.2 to determine the sensitivity of a section to either type of bandwidth.

Example 8.1

Compute the bandwidth of an installed 2-km section with an LED source. Two 1-km sections are used. The following parameters apply:

$$a = 0.71 \qquad \lambda_c = 1320 \text{ nm}$$
$$BW_f = 500 \text{ MHz·km} \qquad \lambda_w = 150 \text{ nm}$$
$$\gamma_{laser} = 0.7 \qquad b_1 = 1.10$$
$$b = 0.25 \qquad b_2 = 0.0189$$
$$\gamma_{chrom} = 0.69 \qquad b_3 = 1370 \text{ nm}$$

Solution Because the transmitter uses an LED source, Equation (8.14) shows that the bandwidth includes both laser and chromatic components. The laser bandwidth is given by Equation (8.13).

$$BW_{s(laser)}(MHz) = 0.71 BW_f(MHz \cdot km) l_s^{-0.7} l_c^{0.25}$$
$$= (0.71)(500)(2^{-0.7})(1^{-0.25}) = 219 \text{ MHz}$$

From Table 8.2, the equation for chromatic bandwidth is

$$BW_{s(chrom)}(MHz) = \frac{10^4 \, l_s^{-0.69}}{\sqrt{\lambda_w}[1.10 - 18.9 \times 10^{-3}|\lambda_c - 1370|]}$$

$$= \frac{10^4 \, (2^{-0.69})}{\sqrt{150}[1.10 - 18.9 \times 10^{-3}|1320 - 1370|]}$$

$$= 247 \text{ MHz}$$

Using Equation (8.14), we obtain

$$BW_{s(LED)} = \frac{BW_{s(chrom)} \, BW_{s(laser)}}{\sqrt{BW^2_{s(chrom)} + BW^2_{s(laser)}}}$$

$$= \frac{(247)(219)}{\sqrt{247^2 + 219^2}} = 164 \text{ MHz}$$

The installed system bandwidth with an LED source is 164 MHz. This illustrates the importance of selecting a fiber at the spooled stage that will ensure the necessary installed section bandwidth.

8.3 OPTICAL MARGIN MEASUREMENTS

An *optical margin* is an end-to-end optic system measurement. Optical margin is that level of received power in decibels, which is in excess of that required to obtain the desired minimum system-performance level. For an analog system this is the signal-to-noise ratio, or in the case of FM, the carrier-to-noise ratio. For a digital system, this performance level is often the bit error rate.

An excessive optical margin on systems often requires the addition of fixed optical attenuators in line with the fiber link in order to prevent receiver overload. The proper attenuation is determined by first measuring the system's optical margin. Refer to Figure 8.9. An optical margin is measured by inserting an optical attenuator in line with the fiber transmission path. The attenuation is measured while the BER is observed. When the error rate increases to the minimum required system level, the additional loss of the optical attenuator (including its insertion loss) is recorded as the optical margin. The optical attenuator can be placed at any point along the fiber link. There is an advantage, however, in placing it near the receiver. This permits varying the attenuation while directly observing the bit-error-rate receiver.

An optical margin test configuration for a typical high-speed, single-mode, fiber telecommunications link is illustrated in Figure 8.10. The optical attenuator is placed between the cable interface box and the optical receiver. The original cable interface to the receiver jumper now connects to the attenuator's input. A BER transmitter is connected to one of the DS3 multiplexer input channels while a BER receiver is connected to the same DS3 channel on the output end of the demultiplexer (see Chapter 5).

Optical attenuators can also be used to simulate the loss of fiber cables and components. First adjust the attenuator for the expected loss. Measuring the optical margin on simulated fibers is accomlished by increasing the inserted attenuation farther while observing the error rate (Figure 8.9). In this measurement, the optical margin is the difference between the loss initially selected for the cable simulation and the total loss now set into the attenuator. Optical attenuators having a relative display mode permit zeroing the displayed attenuation for the simulated cable so that the additional loss inserted (the optical margin) is now displayed directly.

8.3.1 Bit-Error-Rate Measurement

Bit-error-rate tests are measurements of the system performance. These tests indicate performance of the entire circuit from input to output and include both the electrical and optical circuits. The measurements are performed in a similar manner whether the transmission medium is wire, coax, or fiber. For bit-error measurements, a fixed repeating pseudo-random data pattern is generated from a test set (usually called a BER analyzer) and transmitted over the system. The transmitted pattern may be looped back from the far end into the test set's pattern receiver. The test can then detect and compare the received pattern, bit for bit, against the transmitted pattern. Alternatively, a BER pattern receiver located at the far end can detect the pattern and compare against an internally generated pattern. Standards set by AT&T and CCITT define the specific pseudo-random pattern to be used. These pseudo-random patterns are usually referred to as *quasi-random-signal-source* (QRSS) patterns.

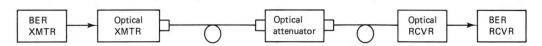

Figure 8.9 Basic optical margin test configuration.

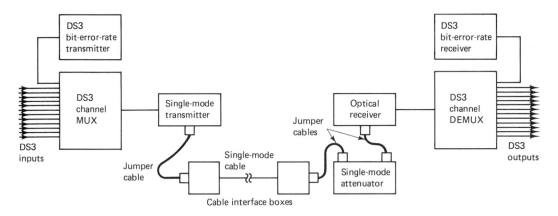

Figure 8.10 Optical margin testing of a typical single-mode DS3 multiplexed fiber system.

8.4 FAILURE TYPES

Failures can be either catastrophic or intermittent. Catastrophic failures are easiest to localize and repair. Troubleshooting generally involves finding the component or subsystem that failed and then replacing it. Components or subsystems that cause degraded or intermittent operation can be more difficult to identify and repair. When a fiber optic system goes down, one of the first areas to check is optical power. Refer to Figure 8.11. Use an optical power meter and check the optical transmitter output at point D. If the power is found to be inadequate, the problem is related to the transmit portion of the circuit. Remeasure the optical power at point C. If there is inadequate power at point C, use an oscilloscope at point B to verify that the transmitter is receiving a signal. This procedure will enable pinpointing the problem.

Tests can be run on the cable plant from D to F with several instruments. A

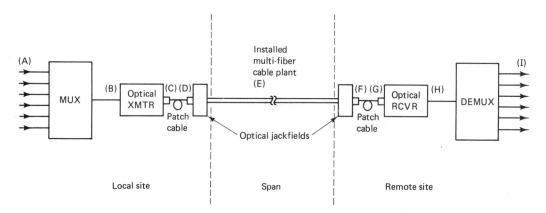

Figure 8.11 Typical fiber optic transmission system (50 percent of a full duplex system).

loss set can be used (see Figure 8.2) or an OTDR can be used to check the attenuation. If the length is very long, the OTDR may not have an adequate range to test the fiber end to end. If the measured loss is infinite or in excess of specification, there is damage somewhere along the span of the cable. The OTDR is useful in locating the fault in terms of distance along the cable.

Failures that are the result of intermittents or degradation require a somewhat different troubleshooting approach. Generally a more qualitative approach is necessary. The performance of the entire system along with the subsystem performance must be measured and evaluated to locate the problem. For example, an end-to-end test for the system would be a bit-error-rate test. For the optical subsystem C to G both optical power and optical loss measurements should be made. This will determine if the optical power is too low or too high. If the optical level is too high, an attenuator can be inserted in line. It is often useful to measure the system's optical margin.

In single-mode systems, if the transmission wavelength is near or outside the fiber's normal dispersion window, excessive dispersion or bandwidth limiting of the signal is probably the trouble. The transmitter wavelength can be checked with an optical wavemeter or optical spectrum analyzer. If this condition is detected, replacing the transmitter should correct the problem.

Additional equipment is available for more specialized tests. Chromatic dispersion and bandwidth test sets can be used to evaluate the information-carrying capacity on both single-mode and multimode fibers. The peak wavelength of the transmitter can be checked by using an optical wavemeter. Problems that seem to defy explanation will sometimes occur. A careful application of these basic tests will usually provide a clue to the problem.

REFERENCES

8.1. JIBBE, M. K., and FURRY, M. E. 1986. Fiber optic testing for end-users. *Proceedings Tenth Annual International Fiber Optic Communications and Local Area Networks Exposition,* 123–131.

8.2. HECHT, J. 1983. Fiberoptic test equipment survey. *Laser Appl.* 2:63.

8.3. DANIELSON, B. L. 1985. Optical time domain reflectometer specifications and performance testing. *Applied Optics,* 2321.

8.4. FRANZEN, D. L. 1986. Standard measurement procedures for charaterizing single-mode optical fibers. *Test and Measurement World,* 70–79.

PROBLEMS

8.1. Discuss the differences between modal, material, waveguide, and chromatic dispersion.

8.2. In Example 8.1, the use of an LED source instead of a laser reduces the installed bandwidth from 247 MHz to 164 MHz. Why would the use of an LED be considered?

8.3. Repeat Example 8.1 for:
  (a) An installed 4-km section, using four 1-km sections.
  (b) An installed 4-km section, using two 2-km sections.
  (c) An installed 8-km section, using 1-km sections.

8.4. Figure 8.7 illustrates obtaining fiber bandwidth directly using a modulated frequency. Devise a test setup to obtain this measurement. Specify the type of modulation, along with the quantity (device) actually being modulated.

8.5. What disadvantage do you see in using the "cut-back" method to measure fiber attenuation?

8.6. In an installed fiber optic system designed with a laser transmitter, what would be the most likely effect of mistakenly replacing the laser diode with an LED?

8.7. Why is the loss of a fiber greater near the transmit end of the fiber than it is farther along the fiber?

8.8. Explain why, in an OTDR, the refractive index set into the instrument must be the same as the effective refractive index of the fiber.

Appendix: Ray Theory

Optical phenomena are often adequately explained by considering light as narrow rays. Based on this approach, this theory is called *geometrical optics*. In the following we summarize the rules associated with this theory.

A. In a vacuum, light rays travel at a velocity $c = 3 \times 10^8$ meters/second. In other mediums, the light rays travel at a slower rate, v, given by

$$v = \frac{c}{n}$$

where n is called the *index of refraction*. Thus, rays travel in straight lines unless deflected by a change in the medium.

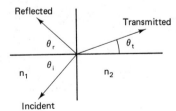

Figure A.1 Incident, transmitted, and reflected waves at a plane boundary.

TABLE A.1 COMPLEMENTARY ERROR FUNCTION TABULATED VALUES

x	10 log x	erfc(x)	x	10 log x	erfc(x)	x	10 log x	erfc(x)
3.10	4.91	9.68E-04	4.10	6.13	2.07E-05	5.10	7.08	1.70E-07
3.15	4.98	8.16E-04	4.15	6.18	1.66E-05	5.15	7.12	1.30E-07
3.20	5.05	6.87E-04	4.20	6.23	1.33E-05	5.20	7.16	9.96E-08
3.25	5.12	5.77E-04	4.25	6.28	1.07E-05	5.25	7.20	7.61E-08
3.30	5.19	4.83E-04	4.30	6.33	8.54E-06	5.30	7.24	5.79E-08
3.35	5.25	4.04E-04	4.35	6.38	6.81E-06	5.35	7.28	4.40E-08
3.40	5.31	3.37E-04	4.40	6.43	5.41E-06	5.40	7.32	3.33E-08
3.45	5.38	2.80E-04	4.45	6.48	4.29E-06	5.45	7.36	2.52E-08
3.50	5.44	2.33E-04	4.50	6.53	3.40E-06	5.50	7.40	1.90E-08
3.55	5.50	1.93E-04	4.55	6.58	2.68E-06	5.55	7.44	1.43E-08
3.60	5.56	1.59E-04	4.60	6.63	2.11E-06	5.60	7.48	1.07E-08
3.65	5.62	1.31E-04	4.65	6.67	1.66E-06	5.65	7.52	8.03E-09
3.70	5.68	1.08E-04	4.70	6.72	1.30E-06	5.70	7.56	6.00E-09
3.75	5.74	8.84E-05	4.75	6.77	1.02E-06	5.75	7.60	4.47E-09
3.80	5.80	7.23E-05	4.80	6.81	7.93E-07	5.80	7.63	3.32E-09
3.85	5.85	5.91E-05	4.85	6.86	6.17E-07	5.85	7.67	2.46E-09
3.90	5.91	4.81E-05	4.90	6.90	4.79E-07	5.90	7.71	1.82E-09
3.95	5.97	3.91E-05	4.95	6.95	3.71E-07	5.95	7.75	1.34E-09

B. At a plane boundary between two mediums, a light ray is reflected at an angle equal to the angle of incidence (see Figure A.1). Thus $\theta_i = \theta_r$, where θ_i is the angle of incidence and θ_r is the reflected angle.

C. If light (power) crosses the boundary, the direction of the transmitted ray is given by Snell's law. That is,

$$\frac{\sin \theta_t}{\sin \theta_i} = \frac{n_1}{n_2}$$

where θ_t is the angle of transmission and n_1 and n_2 are the refractive indices of the incident and transmission regions, respectively.

Index

A

Absorption (*see* Attenuation)
Absorption coefficient, 169
Acceptance angle, 4, 13-14 (*see also* Numerical aperture)
Amplifier noise:
   noise, 47-51, 173-176
   noise figure, 59-60
Analog modulation, 25, 33-34, 43-76
Analog system design:
   bandwidth, 51-59, 67-68, 217
   carrier-to-noise ratio, 59-61, 65-66
APD (*see* Avalanche photodiode)
Attenuation in fibers:
   attenuation, 1, 81
   material dispersion, 8-9
   modal dispersion, 6, 97, 215-218
   Rayleigh scattering, 82, 213
   wave guide dispersion, 8-9

Avalanche photodiode, 27, 40, 54-56, 172

B

Bandwidth:
   analog systems, 53-58, 168-169
   budget, 96-100
   digital data, 121-125
   efficiency plane, 117-119
   laser diode modulation bandwidth, 23-25
   manufacturer's specification, 219-220
   rise time, 39, 72-73, 165-171
   single mode fiber, 67-68
Baud, 129
BER (*see* Bit error rate)
Binary codes, 87-89, 125-141

Bit, 116–117
Bit error rate, 95, 114, 223 (*see also* Probability of error)
Bit time, 97
Boltzmann's constant, 48
Burrus coupler, 21

C

Channel capacity:
  digital, 116–120
  optical, 43
Circuits (*see* Receiving circuits; Transmitting circuits)
Coding principles, 120–121
Coherent optical communications, 46, 187–208
Connectors:
  losses, 31, 91, 105–112
  star coupler, 32–33
  T-coupler, 32–33, 189
Coupling loss, 12–20
Critical (minimum) angle, 4

D

Dark current, 174
Data bus topology, 104–112
Data encoding for fiber optics, 131–141
Depletion region, 164–167
Detectors (*see* Photodetectors)
Diffraction, index of, 2–3
Diffusion time, 167
Digital coding, 87–89 (*see also* Binary codes)
Digital signaling techniques, 125–131
Digital system design:
  alternative design approaches, 70–72
  power budget, 37–39, 91–92
  receiver sensitivity, 65, 82, 94, 196
  required optical power, 94–96

Diodes (*see* Laser diode; Light-emitting diodes; Photodetectors)
Dispersion:
  dispersion shifted fiber, 84–86
  material and waveguide dispersion, 8–9
  modal dispersion, 4, 9–10
  zero-dispersion wavelength, 9
Distortion, 68–70
Divergence, 43

E

Electric field, 46
Electro-acoustic modulation, 197
Electro-optic modulation, 193
Energy per bit, 117–119
Error detection/correction, 119–120
Error rates (*see* Bit error rate)
Excess loss, 31–32

F

FDM (*see* Frequency division multiplexing)
FET amplifiers (*see* Receiving circuits)
Fiber cables (*see* Optic fibers)
Fiber communication systems:
  history, 1
  major components, 2
Filter, 181
Frequency division multiplexing, 57–59
Frequency modulation, 174
FSK encoder, 134–135

G

GaAlAs, 20, 163
Germanium, 176
Glass, 30

Graded index fiber
  index profile, 9–10
  modes, 6
  numerical aperture, 13

H

Heterodyne detection, 46, 187–209
Heterodyne phase locked loops, 205–208
Homodyne detector, 197–205

I

Impulse response, 10–12, 180
Index of refraction, 2–3 (*see also* Appendix)
Index profile, 3–10, 215–217
Information rate limit (*see* Bandwidth; Channel capacity)
InGaAsP/InP, 22, 176
Injection laser diode, 22, 29
Integrated detector preamplifier, 41
Integrating amplifier, 172–173
Intensity (*see* Electric field)
Intensity modulation, 25 (*see also* Analog modulation)
Intersymbol interference, 177–178

J

Junction capacitance, 169

L

Lambertian beam:
  coupling efficiency, 18
  definition, 17
Laser diode, 22–25

Lasing threshold current, 22
Light-emitting diode (LED):
  construction, 20–22
  modulation, 22
  radiation pattern, 17–18
  types, 20
Light reception, 28
Line width, 187–203
Local area networks, 104–112
Loss budgeting in system design, 37–39 (*see also* Attenuation in fibers)
Loss measurement, 211–213
Lorentzian, 203

M

Material dispersion (*see* Dispersion)
Modes:
  degenerate, 8
  LED-based multimode fiber systems, 220–222
  number of modes, 5–6, 8
  single mode, 6–8, 225
Modulation (*see also* Analog modulation; Digital coding):
  multichannel pulse frequency modulation, 75–76
  multiple channel modulation, 57–59
  pulse code modulation, 123–125
  pulse modulation, 72
  pulse frequency modulation, 73–75
  rate, 125
  techniques, 193–195
Multiplexing:
  frequency, 61
  signal, 32
  wavelength division multiplexing, 76

N

Noise:
  Gaussian, 201–202
  phase, 201

Noise (cont'd.)
   shot noise, 45, 49–50
   thermal noise, 47–50
Noise equivalent power (NEP), 28, 169, 171, 175–176
Normalized frequency (V parameter), 5–6
NRZ signals (see Binary codes)
NTSC video, 146–157
Numerical aperture (see Graded index fiber)
Nyquist's rate, 123–124
Nyquist's theorem, 177–183

O

Optic cables, 29–30
Optical channel, 43
Optical margin measurements, 222
Optical radiation, 44
Optical transmission of video, 143–159
Optic detectors (see Photodetectors)
Optic fiber cables (see Optic fibers)
Optic heterodyne detection (see Coherent optical communications)
Optic multiplexing (see Multiplexing)
Optic sources (see Laser diode; Light-emitting diode)
Optic time domain reflectometer (OTDR), 213–214

P

PCM (see Modulation)
Pigtail, 13
Phase locked loop, 204–205
Phase noise, 201–205
Phase-shift keying, 132 (see also Binary codes)
Photodetectors:
   avalanche photodiode (APD), 25–28, 54–59, 198
   PIN diode, 25–32, 162, 199

Photons, 23, 44, 208
Planck's constant, 45
Polarization maintaining fiber, 194–196
Power budget, 91 (see Analog system design; Digital system design)
Probability of error, 95, 118 (see also Bit error rate)
Pulse code modulation (see Modulation)
Pulse frequency modulation (see Modulation)

Q

Quantum efficiency, 23, 45
Quantum limited detector, 84
Quantum limited SNR, 65–66, 200

R

Radiation patterns:
   edge emitter, 26
   fiber, 29
   Lambertian, 17
Rayleigh scattering (see Attenuation in fibers)
Ray theory, 4
Receivers (see Receiving circuits)
Receiving circuits, 40, 62, 71, 173–183
Reflection loss, 18–19
Refractive index (see Index of refraction)
Resonant cavity, 203
Responsivity, 26, 65–66
Rise time (see Bandwidth)

S

Sampling theorem, 124
Shannon's theorem, 115–119
Shot noise (see Noise)

Signal-to-noise ratio (SNR), 52–58
Single mode fiber (*see* Modes)
Source coupling, 18–20
Sources (*see* Laser diode; Light-emitting diode)
Spectral width (*see* Line width)
Spectrum (*see* Impulse response)
Star coupler, 31, 106–112
Star network, 107
Step index fiber (*see* Index profile)

T

T coupler, 31
Transmission format:
   T-1, 146–147
   DS1, 146–147
   DS3, 146–147

Transimpedance amplifier (*see* Receiving circuits)
Transmitting circuits, 40, 61, 63

V

Velocity of light, 2–3
Video:
   analog-to-digital conversion, 151–154
   compression, 154–156
   digital video, 144–154
   NTSC video, 146–157

W

Waveguide dispersion (*see* Dispersion)
Wavelength-division-multiplexing (WDM) (*see* Multiplexing)